Seven Wonders of the Universe
That You Probably Took for Granted

Seven Wonders of the Universe

That You Probably Took for Granted

C. Renée James

with illustrations by

Lee Jamison

The Johns Hopkins University Press

Baltimore

© 2011 The Johns Hopkins University Press
All rights reserved. Published 2011
Printed in the United States of America on acid-free paper
2 4 6 8 9 7 5 3 1

The Johns Hopkins University Press
2715 North Charles Street
Baltimore, Maryland 21218-4363
www.press.jhu.edu

Library of Congress Cataloging-in-Publication Data

James, C. Renée.
Seven wonders of the universe that you probably took for granted /
C. Renée James ; with Illustrations by Lee Jamison.
p. cm.
Includes bibliographical references and index.
ISBN-13: 978-0-8018-9797-9 (hardcover : alk. paper)
ISBN-10: 0-8018-9797-1 (hardcover : alk. paper)
ISBN-13: 978-0-8018-9798-6 (pbk. : alk. paper)
ISBN-10: 0-8018-9798-X (pbk. : alk. paper)
1. Cosmology—Miscellanea. 2. Astronomy—Miscellanea.
3. Physics—Miscellanea. I. Jamison, Lee. II. Title.
QB982.J36 2011
520—dc22   2010015271

A catalog record for this book is available from the British Library.

Special discounts are available for bulk purchases of this book. For more
information, please contact Special Sales at 410-516-6936
or specialsales@press.jhu.edu.

The Johns Hopkins University Press uses environmentally friendly book
materials, including recycled text paper that is composed of at least
30 percent post-consumer waste, whenever possible. All of our book
papers are acid-free, and our jackets and covers are printed on
paper with recycled content.

In loving memory of Wynn Martin,
who taught me that even slugs are wondrous.

CRJ

To Elsa Emmerich Jamison, who taught me how to think
and how to embrace being nuts;

To Melinda Olson Jamison, who could somehow live
and raise children with me;

To Kathy Troquille, the quadriplegic for whom I learned
to paint with my teeth and from whom I learned
where I live.

LEJ

Contents

Preface　*xi*

Acknowledgments　*xiii*

Introduction　1

CHAPTER 1. Night　6

1.1 Of Snowballs and Ice-Skaters　8

1.2 Putting the Brakes On　16

1.3 Why Is Space Dark? Answer #1: Location, Location, Location　23

1.4 Why Is Space Dark? Answer #2: Because　29

1.5 Why Is Space Dark? Answer #3: Actually It Isn't　33

Small Wonder: Day and Night on Mercury　38

Small Wonder: Keeping the Night Sky Dark　40

CHAPTER 2. Light　42

2.1 Codebreaking Basics　43

2.2 The Little Orphan Annie Secret Decoder CD　47

2.3 More Than Meets the Eye　52

2.4 Evading the Question　63

2.5 Making Light of the Universe　69

Small Wonder: Why Is the Sky Blue? And Why Are Sunsets Red? 72

Small Wonder: The Early Universe—A Made-for-TV Movie! 75

CHAPTER 3. Stuff 76

3.1 It's Element-ary 77

3.2 Element Factories 79

3.3 In the Beginning . . . 91

3.4 Making Light of Evil Twins 97

3.5 The Good Guys Always Win . . . But Why? 99

Small Wonder: What Are You Really Made Of? 101

CHAPTER 4. Gravity 103

4.1 A Penny for Your Thoughts? 104

4.2 Earth, the 6 Trillion Trillion Kilogram Weakling 107

4.3 Of Apples and Orbits and Confused Astronauts 116

4.4 But Wait . . . There's More! 121

4.5 Surprise! Gravity Isn't a Force! 124

Small Wonder: How to Lose Weight and Keep It Off, Guaranteed! 133

CHAPTER 5. Time 135

5.1 Got a Second? 136

5.2 Time in a Bottle—or in a Black Hole 146

5.3 A One-Way Ticket to Disorder 154

5.4 Time: The Grand Illusion? 158

Small Wonder: So What Causes Daylight Saving Time to Happen? 161

Small Wonder: Why Are There Seven Days in a Week? 161

Small Wonder: What Would Happen if You Fell into a Black Hole? 164

CHAPTER 6. Home 166

6.1 Goldilocks and the Three Planets 167

6.2 Goldilocks and the Three Stars 178

6.3 Goldilocks in Suburbia, Traffic, and Time 182

6.4 Why Goldilocks Could Never Have Been a Successful Real Estate Agent 187

Small Wonder: Looking for a Home Away from Home 193

Small Wonder: Do Aliens Exist? 195

Small Wonder: No, Really. Do Aliens Exist? 195

Small Wonder: Wanted—Alien Hunters 196

CHAPTER 7. Wonder 193

7.1 Life—The Ultimate Eating Machine 199

7.2 There's More to Life Than Eating 206

7.3 It's All in Your Head . . . Maybe 214

7.4 It's a Wonder-Full Life 222

Small Wonder: Small Wonders 224

Afterword 228

For Further Reading 229

Index 235

Preface

You've probably picked up this book because you wonder what it's like. You wonder if the science is going to be too difficult and boring, or too easy and boring, or at a good level for you, academically speaking, but still boring. You wonder if the style will appeal to you or your kid or your spouse or your friend for whom you desperately need to buy a gift sometime before the party at six o'clock. You wonder if there's an appropriate card and gift bag to match, and whether you have time to grab something from the café. You wonder, as you absentmindedly flip through the pages, how there can even be a book about the wonders of the Universe that isn't packed with awesome full-color photos, instead sporting some bizarre sketchy cartoons.

You have not yet begun to read about the wonders of the Universe, but you've managed to wonder about quite a few things already. And that's fantastic because that means that you are the exact person I had in mind while writing this book.

You might wonder how such a book got its start. Well, I'll tell you. The idea was hatched not long after the 2007 worldwide vote on the Seven Wonders of the World. Tragically, the giant Sam Houston statue on Interstate 45 near my hometown of Huntsville, Texas, did not make the list, so I had to try something different, something bigger. Forget the wonders of the world. I wanted to tackle the wonders of the entire Universe! The problem I ran into was that there are so darned many gorgeous photos of things in the Universe and very few reasonable guidelines for whittling that list down to some arbitrary number like seven. Why not ten? Why not seven hundred? But more important, what publisher would want to produce a book full of photos that you can download off the Hubble Space Telescope website?

So I began to think about less photogenic things. Everybody loves a great spiral galaxy, but nobody appreciates the wonder of gravity that made the galaxy possible in the first place. If you suddenly found yourself cringing when you saw the word *gravity*, this probably means that you've had a class where you were fed the dry facts and Newton's equation, but never the wonder of it. It's pretty mystifying, actually. More so than Saran Wrap, which, according to at least one person out there, should have been included in this book.

Now, I'm not suggesting that Newton's gravitational equation isn't incredibly useful. It's something that bridge builders and aerospace engineers Really Need to Know, and it's arguably more important than Saran Wrap. But this isn't a textbook. The discussions in here aren't even remotely complete. Entire scientific careers are devoted to things that I gloss over with a sentence fragment or, worse yet, ignore completely. Sure, it's partly because I am totally unaware of these subjects, but it's also partly because the book would have gone quickly from an inviting little tome that my kid's English teacher could read in a weekend to a ginormous set of encyclopedias that nobody would ever bother opening (but would likely brag about owning). In fact, I'm sure my biologist friends would be mortified to find that I never mention DNA, the molecule responsible for encoding all of life on Earth. Never! (Well, assuming you don't count this mention of it.)

That's because this book is not supposed to be a comprehensive dissertation on anything. It's just meant to get you to look at the Universe you inhabit with the same sense of wonder that a small child has.

I wonder if it will.

Acknowledgments

It's really no wonder that this book couldn't have come together without the help of a vast army of people. Obviously, there's Johannes Gutenberg, without whose invention of the printing press, things like mass-produced books would never have become commonplace. Thanks, Johnny.

Then there are my parents, without whom I wouldn't be doing, well, *anything*. In addition to giving me life, they did a great job of encouraging me to think critically about the Universe while still fostering a playful sense of wonder (once I got past those sullen teenage years, that is). I am particularly grateful that they decided to tune into PBS when I was in sixth grade so that I could witness the beauty of Carl Sagan's *Cosmos* series. That, probably more than any other outside influence, drove me to seek the wonder in everyday experiences.

Even closer to home—actually, in my home—are several people who have contributed one way or another: my husband, Sam, who helped me realize that my "lucid" explanations weren't always so lucid to people who aren't trained scientists; my sixth-grade son, Sean, who kept assuring me that everything was "cool, really cool"; my two-year-old daughter, Megan, who often tested my ability to see wonders amid everyday life; and my infant son, Jamie Keegan, who, because of his birth halfway into the writing of this book, is responsible for any incoherent sentences you might encounter. Seriously, though, these people are a wonder, and whether they know it or not, they continually help me see the world through fresh eyes.

Because my formal training is in astronomy, I had to pick the brains of experts in other fields that are featured in this book, notably geology, biology, and philosophy. Dr. Brian Cooper's introductory geology notes helped with the first subject, while Dr. Frank Fair provided a veritable summer reading list in philosophy, which was distilled into three paragraphs in the discussion of "consciousness." Dr. Christopher Randle was kind enough to provide some up-to-date references on the emergence of life from a biochemistry standpoint, information that again became concentrated like the core of a massive star just before it goes supernova. In other words, it was squished down a lot, and I leave it as an exercise for the reader to find out just how much.

Then there are the folks who agreed—for whatever bizarre reason—to read every single chapter of this book as it was written and comment on both the style and content: Dr. Derek Wills, Dr. John Beam, and Lee Jamison. Not that the last one could have gotten away without reading it, because being familiar with the entire book was sort of a prerequisite for doing the artwork. Still, his nonscientist insights were quite valuable, as were the sleep-deprived stream-of-consciousness meetings about what concepts should be illustrated and how. The omnipresent penguins were his idea.

Several other individuals were subjected to only portions of the book, and again provided valuable feedback: Dr. Brian Cooper, Dr. Frank Fair, Dr. Barry Friedman, Dr. Pamela Gay, Brian Gedelian, Dr. Mary Kay Hemenway, Father Stephen Payne, and Michael Prokosch.

There were hordes of people (specifically, two hordes) who were willing to share their list of wonders with me. The first horde consisted of astronomy educators (formal and informal) around the globe who, when the book was still in its larval phase, suggested several topics and anecdotes that ultimately made it into the book. They are (in alphabetical order): Richard Alvidrez, Philip Blanco, Karen Castle, Erin Dokter, Joann Eisberg, Jerry Fuller, Harold Geller, Michael Hamlin, Adam Gabriel Jensen, Pat Keeney, Liam McDaid, Aileen O'Donoghue, Elisha Polomski, Chris Sirola, Angela Speck, Alex Storrs, Kristine Washburn, Pam Wolff, and Mike Zawaski.

The second horde consisted largely of nonscientists young and old whose "wonders" didn't fit very neatly into the overall narrative of the book, so I simply listed their wonders in the last chapter. This horde included Jan Arnopolin, Shelly Butcher, Karla Christian, Liz Conces, Linda Drewry, Ron Griffiths, Susan Hightower, Janice Jordan, Dr. Peter Kreeft, Scott Martin, Geri Peak, Gwen Penny, Suzi Skutley, Jennifer Stilwill, Christopher Vickery, and Kwenton Williams.

Many thanks also to Emo Philips and Kipleigh Brown, whose cat story provided the inspiration for the fat cat's gravitational well (you'll see what I mean in chapter 4).

Finally, I would like to thank the editors at the Johns Hopkins University Press. First there was Trevor Lipscombe, who agreed to take on this project despite its complete lack of formal scholarship, footnoting, and detailed bibliography. Then there was Mary Lou "Red Pen" Kenney, who did her level best to let me keep the informal style while still adhering to the rules of grammar (and who pointed out a dangling element for which I would have received a C in Mrs. Bigbee's class). It's been a wonder-full experience, one I hope to share again with these folks.

Introduction

There are two ways to live your life—
one is as though nothing is a miracle,
the other is as though everything is a miracle.
—Albert Einstein

It's easy to take Einstein's words to heart when you're riding in a gondola lift toward a sun-splashed peak in the mountains. Surrounded by miles of beauty in every direction, breathing the fresh mountain air. Aaaaaah . . . clearly miracles are everywhere, and you wonder how life in such a place could ever be anything but endless awe. Fantasies of quitting your job and washing tour buses just for the privilege of living there roll through your mind as the cable car climbs toward the summit. A passenger next to you voices the exact thoughts you've been having, when suddenly your dreamy trance is shattered by the teenaged tour guide.

"Ya get used to it," he says.

It's the same tone of voice a veteran parent might use when discussing diaper changes.

'Used to it?' your brain protests wildly. 'Are you nuts?'

But because you need him to guide you back off the mountain after you take lots of pictures, you don't say this out loud.

It's a mixed blessing that humans are amazingly good at getting used to things. The same ability that allows us to adapt to the bad times also makes us ignore the good ones after a while. Even in the unparalleled beauty of the mountains. Even in a city like Rome, filled to the brim with history and culture.

Imagine! Thousands of Romans stream past the Coliseum every day, giving it scarcely a thought. To them, it's just another structure in their hometown, made inconvenient by the throngs of tourists that keep clogging the traffic in that area. To the rest of the world, the Coliseum is a wonder of architecture and history, and we'll even pay good money to get a t-shirt to prove it.

Chances are you don't live near any of the officially recognized wonders of the world. But amazingly enough, you can witness the seven wonders of the Universe from your own home, even if you're engaged in something less than wondrous.

"Really?" you ask eagerly after a dreary day of work and housecleaning. "Tell me how!"

Okay. Picture this:

It's Wednesday night. Late. And you have just remembered that tomorrow is trash day. So you begrudgingly drag yourself into the frigid night air as thousands of stars shine mockingly down at you. You ignore their cold stares and head to the garage, where you find that someone has already tried taking the cans to the curb for you. Unfortunately, that someone was a raccoon, and he didn't do a very good job. The unkind words you have for the raccoon come out in little humid puffs as you set the cans upright again and try to replace their contents using only your thumb and forefinger. Victorious, you finally wrestle the fully contained trash to the curb, only to notice the rising half Moon. In the deep recesses of your sleepy, irritated brain is the knowledge that a rising half Moon means that it must be around midnight.

And tomorrow's the Big Meeting.

"Just my luck," you mutter, and you trudge defeated back into the house.

There you go! All seven wonders of the Universe, and you got to experience them in a span of five minutes.

"Um . . . perhaps I'm missing something, but that didn't seem all that wondrous."

Sure it did. But in case you didn't catch them, here's the slow-motion replay:

It's Wednesday night.

(*Night! Wow, what an amazing concept! Here you are, on a planet that shows one face to the nearby bright Sun, and then points it outward to the rest of the much-darker Universe. This isn't something to be taken for granted. Just look at the Moon, perpetually pointing the same face to Earth. On another planet around another star we, too, might have been tidally locked to our Sun, never experiencing night without a pilgrimage to the "dark side." But even on that planet, the "dark side" requires . . . well . . . dark. Cosmically speaking, even this phenomenon isn't a given. In another part of the Universe, or even in an alternate universe, the sky we see when we're not facing the Sun might be blazingly bright. You eagerly head to the door to immerse yourself in this phenomenon called night.*)

Late.

(*So it's late, eh? What exactly does that mean? It's no longer "early" by your standards, so you apparently have some way of measuring this concept we call time. As long as there have been poets and philosophers and physicists, people have puzzled over the nature of time. And here you are, experiencing it in all its glory. Lateness! Another amazing thing!*)

And you have just remembered that tomorrow is trash day.

(*Remembered? How is that possible? Are you not merely 7 trillion quadrillion atoms, siphoning energy from other atoms that you call food? What really distinguishes you from that refuse you're trying to eliminate? Somehow you have con-*

sciousness. You are, as René Descartes succinctly reasoned, by the fact that you are a thinking being. Wow. A thinking being. Late. At night. As if that weren't enough, you're about to deal with trash! That's pretty unbelievable in itself because, like you, trash is stuff. Protons, neutrons, electrons. Products of billions of years of cosmic evolution, and you are about to interact with it. Maybe you should forget the trash more often.)

So you head into the exhilarating night air where thousands of stars smile down on you.

(Do you mean to say that there are thousands of objects, trillions of miles away, that you are aware of simply because they are shining? Yes! And if you understood how to decode that starshine, you could read messages about their motions and makeup, their masses and magnetism. Even secrets from the very dawn of the Universe are etched in this thing we call light. And, although you can't personally perceive every type of light there is, something about your physiology lets the Universe communicate with you even when you can't touch it. This night just keeps getting better!)

You bask in their faint light and head to the garage, where you find that someone has already tried taking the cans to the curb for you. Incredibly, that someone was a raccoon. The praise you have for the raccoon comes out in little humid puffs as you set the cans upright again and try to replace their contents using only your thumb and forefinger.

(You can barely contain your ecstasy now! In just a few seconds' time, you have encountered two of the greatest wonders of the Universe. One of them is gravity. Not only does it give us a sense of "uprightness," but it's also amazingly democratic, treating all objects the same. Without it, trash cans couldn't be toppled, stars couldn't form, and planets couldn't

*exist. Let's hear it for nocturnal demonstrations of gravity!
What's more, those mischievous little Earth-dwellers that
toppled your trash further remind you of the uniqueness and
wonder of your home planet. Neither too close to nor too far
from its ideal star, neither too large nor too small, your home
planet is a rare—if not unique—place. Amazing!)*

Now that the trash is out and you are graced by the rising
third-quarter Moon (which only inspires more awe as you consider the impact this object has had on our timekeeping), the Big
Meeting doesn't seem quite so daunting.

'What luck!' you think, as you marvel at all the wonders of the
Universe you've just encountered—night, light, stuff, gravity, time,
home, and wonder—and head back inside.

Somewhere in the distance, you hear your neighbor cursing at
the raccoons that are gleefully dancing in his trash cans.

As you can see, Einstein was right. And so was Tom Sawyer, for
that matter. Wonders can be found even in the most disgusting of
chores. Now wash your hands and read on to find out what makes
these things so amazing.

CHAPTER 1

Night

*I often think that the night is more alive
and more richly colored than the day*
—Vincent Van Gogh

A young child can force you to the limits of human knowledge with a surprisingly small number of questions. One of the first and simplest questions a preschooler asks is, Why is it dark at night? Sure, this poser might seem easy enough on the face of it, but the reality is that this is one of the most profound problems ever explored, one that has kept cosmologists occupied for as long as cosmologists have existed.

"So why is it dark at night?" your preschooler insists.

Grateful that she didn't ask you where babies come from, you grab that priceless autographed baseball from the shelf and begin your own "Astronomy in Interpretive Dance" show in your living room. "Pretend the baseball is Earth," you explain as her eyes grow wide with reverence. "Earth is just a giant ball that spins every 24 hours. And that giant ball goes around another, even bigger, shining ball called the Sun."

So far, so good, it seems. "See how Nolan Ryan's autograph moves from the lit-up part to the shadowy part of the ball?" you ask. "That's just like us going from day to night, and it happens every time the ball spins around."

As you demonstrate the situation, a fountain of eager ques-

tions spills forth from the child's mouth: "Why does the baseball spin? What would happen if it didn't spin? Why does that bigger ball shine so brightly? Is it plugged in like the lamp? Why aren't there other things out there that shine so bright? Why is space so black? Who is NoOne Rhyme?"

Baseball still in hand, you stare blankly at the child. "How about some ice cream?" you suggest, and the lesson on Earth's day-night cycle is cut short. But those questions nag at you, keeping you awake during the very night that you're puzzling over.

1.1 Of Snowballs and Ice-Skaters

'Well, it *is* the spin,' you think triumphantly as you toss in bed for the third hour straight. But how it got that motion is no clearer to you than the bizarre dreamlike vision of a gigantic basketball player spinning Earth on his finger, each flick of his astronomically large hand helping to maintain its speed.

The Universe is filled with spinning objects, though, and you know there can't possibly be enough colossal basketball players to go around. The solar system itself has scores of rotating things: pretty much everything from the Sun, a blazing ball in which you could fit a million Earths, all the way down to the puny half-mile-wide asteroids rotates. All the planets (yes, even the recently demoted—er, I mean, *reclassified*—Pluto) rotate. So do the moons around those planets.

But why?

A tiny clue about the underlying source of that rotation can be found by looking at how fast each planet rotates. We think we're pretty zippy here on Earth, whizzing around on our axis every 24 hours. That means that a person standing on our equator goes riding along at a decent clip of 1,000 miles per hour!

Few of us actually live at the equator, though, so we don't make quite that big a circle every 24 hours. In fact, the farther north or south you go, the slower you move. Santa Claus—or anyone else at the North Pole or South Pole—doesn't travel an inch around our axis. All he and the elves do is spin in place as though they're riding on an enormous bar stool. Still, most people live at a latitude that covers a few hundred miles every hour. This is great news for you if someone ever complains that you haven't moved all day. Just politely respond, "Actually I've covered quite a bit of territory, thank you very much."

Of the four inner planets—Mercury, Venus, Earth, and Mars—Earth actually will give you the best carnival ride for your buck, which is pretty convenient because, presumably, you already live here. An hour on Mercury's equator won't get you very far at all, as it takes nearly two Earth months just to make a single rotation on its axis. Given that this is two-thirds the amount of time it takes to get all the way around the Sun, it makes for a really interesting day-night-year cycle (see "Small Wonder: Day and Night on Mercury").

Venus is even worse. It takes about eight Earth months to spin once. *Eight months!* Someone on its equator could actually *walk* fast enough to counteract the measly 4-mile-per-hour motion from its rotation, except for the unfortunate fact that that person would be crushed and melted by the Venusian atmosphere.

Details.

At least Mars is respectable. It maintains a day-night cycle that is nearly identical to ours, tolling its midnight bell every 24 hours and 40 minutes. Because Mars is about half the width of Earth, an equator-dweller there cruises around at 540 miles per hour.

But that's nothing compared to the ride on the solar-system heavyweights Jupiter, Saturn, Uranus, and Neptune. If planets were ski slopes, these guys would be black diamond, while Earth would be a bunny slope. If planets were cars, these guys would be Formula 1 racers, while Earth would be a little remote-controlled car in your kid's bedroom. You get the idea.

Of the four big ones, Jupiter is the reigning champ. Weighing

in at over 300 times the mass of Earth, and so huge that more than 1,300 Earths could fit inside it, Jupiter would seem at first glance to be a bit too hefty to do any quick twirling. But twirl it does. This gargantuan planet makes a complete rotation in just 10 hours, spinning so fast that its equator bulges out noticeably. In fact, Jupiter is nearly an entire Earth-diameter *wider* than it is *tall*.

The story for Saturn is similar. A mammoth of a planet made from 95 Earths' worth of material and big enough to hold 770 of our puny planet, Saturn rotates once in a little over 10 hours. Like its big brother Jupiter, Saturn is wider than it is tall because of its incredibly fast rotation. For Saturn, the problem is even more exaggerated by its overall puffiness. It's fully 10% thicker at the equator than it is at the poles.

So how did these solar system Goliaths get so wound up?

To understand this one, you'll need to entertain your pre-schooler with an interpretive dance involving snowballs and ice-skating.

Imagine some kids making a snowman. Those of us living in warmer climes will have to take the word of a movie that kids actually can make person-sized snowmen. The kids begin with a snowball and then roll it through the freshly fallen snow. This sticks to the "starter" ball, making the snowball bigger. Bigger snowballs make bigger ruts in the snow and collect more snow even more efficiently. Eventually, after enough rolling and snow-sticking, the kids have created a two-foot-tall snowman base.

At least, that's what the movies imply. My kids, on the other hand, have only managed to make snow-mites, but I digress . . .

The thing about snowballs is that without enough available snow and some "sticking" mechanism, the ball will never get larger. Same goes for a planet. And, just as the larger snowball is more effective at scooping out the snow and growing even larger, planets like Jupiter can . . . well . . . keep snowballing as long as there is a good supply of stuff to accumulate.

The primary sticking mechanism for the swirl of gas and dust that made up the early solar system was, of all things, static electricity. Yes, the same thing that makes your socks stick together and adheres packing peanuts to a curious cat actually has the power to pull together a blob of cosmic material up to the size of a large boulder, give or take. After this, other sticking forces begin to take over, and as more stuff sticks, the stronger those sticking forces become, which makes more stuff stick, which makes the sticking forces even stronger, until the cosmic pathway has been pretty well vacuumed. After just a few million years of rolling through the ever-thinning pancake of gas and dust of the early solar system, a monster like Jupiter is born.

Surprisingly, to make a small planet like Mercury or Earth the process takes even longer. The Sun is partly to blame, just as it is when kids want to build a snowman. While it doesn't exactly melt the stuff in the inner part of the solar system, the Sun does tend to kick it outward, leaving little material behind for the little snowballing planets to pick up. Their diminutive masses also make it harder to pick stuff up, so small planets actually wander around for upwards of 100 million years, looking for stray dust and rocks to add to their stores. Meanwhile, the Jupiters and Saturns and Uranuses and Neptunes have already greedily scooped up all the material kicked into their neighborhood by the Sun.

Some planets get all the breaks.

Now, before you complain that none of this will help your preschooler understand why the big ones are spinning so rapidly, remember that your revised interpretive dance involves iceskaters, too. Yes, another winter activity that many of us will witness only on television, and then only if it's an Olympic year.

In addition to having an insanely high resistance to motion sickness, ice-skaters have the keen ability to transform the laws of physics into an awe-inspiring spectacle. One of the classic moves is for a slowly spinning skater to gradually draw her arms inward. As she does, she twirls ever faster until her skirt and

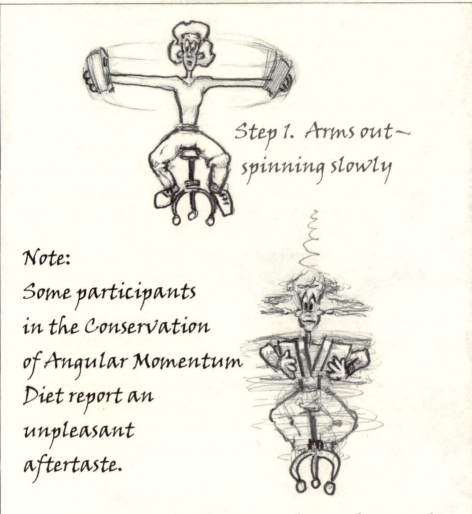

Step 1. Arms out—
spinning slowly

Note:
Some participants
in the Conservation
of Angular Momentum
Diet report an
unpleasant
aftertaste.

Step 2. Arms drawn in—
spinning RAPIDLY

hair are practically horizontal. This effect is due to something
called the conservation of angular momentum, which is a universal law that helps govern the behavior of everything from the
very large (galaxies) to the very small (atoms).

If you don't want to trust the television or attempt the moves
of a professional ice-skater to experience this law for yourself,
you're in luck. All you need are a couple of weights—books, perhaps, or bricks, or even dumbbells—and a chair that spins around,
preferably with little resistance. The little spinning stools in the
doctor's office are great for this, so make an appointment for
that checkup you've been putting off and try this experiment
while you're waiting. Bricks might look a bit out of place in the
waiting room, though, so you might want to take books as your
weights. Nobody would question your ability to read through both
War and Peace and *Gone with the Wind* while waiting for your
appointment.

Once you've been called back to the examination room and
have commandeered the spinning stool, hold your arms out-
stretched with one weight in each hand. Kick yourself into a slow
spin. Try not to level any important medical instruments with
your hands. Now bring your hands toward your chest.

Go ahead. Find some place to do the experiment. I can wait.

[soothing music piped into the examination room]

So did you feel the change?

The best part about doing this experiment at the doctor's office
is that the doctor will be able to dose you up with a nice anti-
nausea medication when you're done. But before your inner ear
begins to protest, you should feel a definite acceleration of your
rotation. The change can be quite dramatic depending on the
weights that you use. Even a gradual, almost imperceptible initial
rotation can get pretty zippy with the right spinning stool and
weight distribution.

Now extend this finding to those snowballing planets. If a few pounds of books can dramatically increase the rotation speed of you, a 100-plus-pound person, simply by moving 20 inches closer to the middle of your chest, imagine how much a gradually tumbling rock will speed up as it pulls trillions of times its own weight toward itself!

This isn't the whole story, of course. Astronomers are pretty sure that this is the process responsible for winding up the big planets like Jupiter, Saturn, Uranus, and Neptune, but scrawny ones like Earth are forced to pick up a rock here, endure a jolting collision there. It's much more like a bumper car ride than the smooth process of drawing material inward, so it gives messier results. Imagine trying the spin-on-the-stool trick at the doctor's office while having some kids run around you and hit your arms occasionally. Still, it seems to explain how the two planets, Venus and Earth, who are practically twins in their masses, sizes, and birthplaces, wound up with such drastically different rotation rates. But it doesn't remotely address how the Sun, an object with a thousand times as much stuff as the rest of the solar system put together, casually spins along at one rotation every 25 Earth days.

Our solar system has been around for a good 4.6 billion years, and in that time plenty of additional factors have complicated matters: friction, heat, turbulence, magnetic fields, and plenty of random collisions just to name a few. Still, we have a good start for the question of why Earth and the other planets spin at all. At the very least it's probably enough to satisfy a few of the child's "whys," and there were no enormous basketball players or ice cream bribes required.

1.2 Putting the Brakes On

Earth itself hasn't always had a 24-hour day, nor will it continue to. In fact, at this point in Earth's history, the length of our day is increasing by about 2.3 milliseconds per century. What this means is that our current day is about 2.3 milliseconds *longer* than it was 100 years ago. For instance, February 23, 2009, was ever so slightly longer than February 23, 1909. February 24, 2009, was ever so slightly longer than February 24, 1909. If you'd had some amazingly precise watch built on February 23, 1909, it would have expected February 23, 2009, to be 2.3 milliseconds shorter than it really was. Same for February 24, 2009. So by the end of the day on February 24, 2009, your antique watch would have accumulated an error of 4.6 milliseconds. And after 1,000 days, your great-grandfather's watch would be a full 2.3 seconds off, so you'd have to adjust it. After all, you wouldn't want to be 2.3 seconds early to a meeting, would you?

If you'd ever wondered about those bizarre "leap seconds" that are announced periodically, this is why they get tossed in. Frustratingly, there are several things that can shorten or lengthen our day unpredictably—earthquakes that adjust the distribution of Earth's mass, internal magma currents, and a host of things that almost nobody ever thinks about—so we don't always know when the next leap second is going to be needed. The last one was inserted during the New Year's countdown between 2008 and 2009. On that day, official clocks read 11:59:59 and then 11:59:60 before hitting the 12:00:00 that signaled the New Year. Most partygoers probably squandered this extra second.

The "standard" day was set about 50 years ago, so we are now essentially looking at a watch that's running a bit too fast for our current day. This is one of the many reasons a body called the International Earth Rotation and Reference Systems Service exists.

Well, that and it's just for the fun of knowing that hordes of scientists around the world spend their ever-lengthening days studying tiny variations in Earth's rotation rate.

If 24 hours seems little enough time for you to accomplish what you need to, imagine how you'd cram all of your activities

into the 4-hour days that Earth was born with. Fortunately over the past 4.6 billion years, outside forces have conspired to drastically slow our rotation rate. And in the coming billions of years, the length of our day will keep increasing until it is about 47 times what it is today.

If you're suddenly thinking, 'Hmm . . . cryogenics,' as you hatch a scheme to enjoy those long, lazy days, think again. The Sun's going to swell up and incinerate Earth long before you'll get the chance.

There's always something . . .

So what is happening to slow our planetary ice-skater? Plenty, as it turns out. Even a real ice-skater can't keep spinning at the same rate forever. For one, the contact between her blades and the ice creates friction that gradually siphons away her rotational energy. Even without that contact, though, eventually the drag between her and the air would slow her down. And just like our ice-skater, Earth is experiencing something called frictional braking. In Earth's case, it's a cosmic collaboration between our oceans and our Moon causing the bulk of the problem.

It all starts with the simple phenomenon that washes away unsuspecting sand castles. If you've ever been to the beach, you might have noticed that the waves will alternate between creeping farther inland, decimating your castles and soaking your beach towels, and receding, leaving behind dead jellyfish, goopy seaweed, and a truly memorable aroma. High tide and low tide, in other words. They happen every day. Nearly two full high-low cycles each day, in fact. And they are evidence of a bizarre force that can act across the vacuum of space.

The idea that the Moon reaches out an invisible hand to distort our oceans was dismissed as preposterous and mythical by intellectual giants such as Galileo, but the fact of the matter is that it does. These forces pull most strongly on the side of the planet that faces the Moon, raising the water levels. They also give a tug to our entire planet, leaving behind the water on the opposite side, so that it, too, is deeper than average.

What the Moon would like to do is keep those two high-water marks—or tidal bulges—precisely in line with itself. The problem is that Earth is spinning blithely away every day, while the Moon adopts a far more casual pace in its orbit around us—it takes a bit over 27 days for a full revolution. As a result, Earth actually pulls those bulges ahead of the Earth-Moon line, while the Moon tries valiantly to pull them back in line. In this way, the Moon is a sort of cosmic cowboy digging his heels into the dirt and screaming, "Whoa, horsey!" as the Earth-horse drags him unceremoniously around the arena.

Outweighed by a factor of 100, the Moon might seem to be exerting this effort in vain, but it is making measurable progress. As Earth tries to keep swimming against the tide, so to speak, friction between the "solid" Earth and the oceans is generated. All in all, about 2.5 trillion watts of power are constantly being removed from the vast storehouse of Earth's rotational energy, mostly in the open ocean as the water rubs against the rugged sea beds. To give you an idea of what this means, it's the equivalent of 5 million high-performance cars all making full use of their 600-horsepower engines in an effort to stop Earth from spinning.

And they're doing this 24/7.

All for the measly result of making our day 2.3 milliseconds longer every century.

To be fair, the Moon isn't trying to *stop* our spinning. It's only trying to get those pesky tidal bulges to stay in line with it. We did the same thing to the more molten Moon billions of years ago, which is why it perpetually shows us the same face. If the Sun had the decency not to vaporize us all in the next few billion years, the Moon would finally succeed in getting one side of Earth to face it perpetually, but by then our days would be over 1,100 hours long.

Perhaps *then* you could do everything you needed to do in a day.

Just the idea of a tidally locked Earth-Moon system is pretty awe-inspiring. Imagine seeing the Moon straight overhead, hour after hour, day after day, never rising or setting unless you personally made a trek around Earth. This is the sort of view astronauts on the Moon have had of their home planet. There's a rather famous picture called "Earthrise" that was taken by the *Apollo 8* astronauts as they went flying by the Moon. If they had actually been standing on the Moon to take the photo, Earth wouldn't have been rising at all. It would just have been hanging there forever.

"This is all really fascinating," you say. "But how does this relate to our experience of night?"

I'm getting there. You see, the Moon isn't the only thing trying to get us in line. The Sun has measurable tidal effects on us as well, but because it's so much more distant than the Moon, the tidal braking we feel from the Sun is much less pronounced. If we were closer to it, it could do a much better job of lengthening our day and forcing us to point one face to it at all times.

Unfortunately, being that much closer to the Sun would be, well, fatal. But not all stars pump out as much energy as the Sun does, and some of those milder stars have planetary systems. Some of the planets in those systems are even hugging pretty close to their stars, and it's highly likely that some of those have even been cozied up to their suns for billions of years. If our scrawny little Moon can rein our rotation in as much as it has over the age of the solar system, imagine what the day might be like on a planet that's tidally locked to its sun. One face always light; one face always dark. No day-night cycle at all.

A being on the day side would never experience night. Never. A creature on the night side would have no clue what high noon meant, and nobody on the planet would know how the heck the sun could appear to rise or set.

If this seems too much like science fiction, consider one of the more recent planetary finds around a star by the name of Gliese 581. Just rolls right off the tongue, doesn't it? Anyway, Gliese 581 is known to have at least three planets in its solar system (probably more), the closest of which would be incineratingly close to the Sun if it were in our solar system. By comparison, Mercury—our swift "messenger of the gods"—would be a frozen wasteland lumbering around the Sun. Fortunately, Gliese 581 is a smaller, cooler star than our Sun, so it's possible to crowd it without being completely vaporized. Although astronomers don't know for certain, it's pretty plausible that all three planets have become tidally locked to their star. There are even some murmurings that at least one of the planets might be capable of sustaining some kind of life.

So imagine that you are that life, and your home planet is Gliese 581d, the furthest of the trio (although you prefer to call it Xchwokdn), and you are a daytime dweller. The concept of a birthday would have absolutely no meaning because if it's high noon when you're born, it's always high noon. In fact, it's always the same "time" of day, as your shadow would never shorten or lengthen. If you could somehow measure the position of your planet relative to your star, you'd find yourself celebrating your birth every two and a half Earth months. Perpetual sunlight would just be a way of life, and, except for whatever Xchwokdnian weather systems might pass over, your view of the sky would never change. Stars? What are those? If you're lucky, perhaps a moon or two is orbiting your planet, so you could at least see *something* change in your sky, but otherwise you're a prisoner to monotony.

Now, imagine that you grew up on the night side. Things would be a bit more fascinating over there. For one, you'd see a full cycle of seasonal constellations rise and set for every trip you made around your sun, so it'd be tons easier to pin down the location of the planet at the time of your birth. To see why this is the case, imagine riding a carousel. A really huge cosmic one, where your sun is the middle post and the rest of your universe is the carnival surrounding the carousel. If you turn to face the middle post (perhaps to stave off motion sickness), your view never changes, just like the case for the residents of the day side of your planet. But if you look outward to the rest of the carnival—to the "night" side—you'll see the popcorn stand whiz by, then the guy selling balloons, then the clown scaring the children away, then the petting zoo. Eventually you come back to the popcorn stand and the cycle begins all over.

Perhaps you were born at the instant that a pattern of stars that look just like a popcorn stand came into view. If so, everyone would celebrate your birth *every* time the constellation Popcorn Stand came into view. "Good ol' Zwkodkb, child of the constellation of Rising Popcorn Stand," they'd call you. Unfortunately, this frequent partying would get old really fast. If you were anything like human (which you most assuredly would *not* be) you'd be over 5 Xchwokdnian years old before you even learned to walk, 10 before you got out of diapers, and kindergarten wouldn't start till you had blown out over 24 candles on your cake. But at least all those candles would help light up the perpetual darkness, which, by their 24th birth party, the Xchwokdnian preschoolers have probably asked about.

"So why *is* the rest of space so dark?" they—and every other preschooler in the Universe—ask of their befuddled parents.

No amount of ice cream will help you bribe your way out of this one.

1.3 Why Is Space Dark? Answer #1: Location, Location, Location

If your preschooler isn't paying too much attention, you could simply try this trick. Have her stand next to a lamp and twirl around. With great confidence say, "You'll notice that when you're facing the lamp, it's much brighter, but when you face away from it, it's much darker. That's because you're really close to the lamp. Likewise, Earth is really close to the Sun. All the other stars are much farther away."

With any luck, she'll be too busy studying the purple streaks in her vision after this demonstration to follow up and you'll be off the hook. Realistically, however, you'll probably hear the next questions: "But aren't there bajillions of other stars? Wouldn't that be like having lots of tiny lamps all over the room? Wouldn't that make the rest of the room look bright, too?"

Deep down you know she's right, and you can't quite figure out why the night doesn't blaze as brightly as the day. There are easily 100 billion stars in our Milky Way Galaxy alone (you might emphasize the name of our home Galaxy just to buy some more time, at which point the child might squeal, "Ooh! Can I have a candy bar?"). And at last count there were over 100 billion known *galaxies*. Seems like plenty of distant points of light to brighten up our night sky. So how is it that they don't make our night sky as painfully bright as our day sky?

Part of our darkness stems from our location. The Sun is essentially in the Galactic suburbs where the population density is pretty low. If you were to make a scale model of the Universe and shrink the Sun to the size of, say, an orange, you'd have to be careful not to step on the puny speck (1/25 of an inch or so) representing Earth 35 feet away. To get to the next nearest orange/ star—you'll need to make your flight reservations. It's fully 1,800 miles away. That's about the distance between Houston, Texas, and San Francisco, California. The traditionally accepted solar system (the Sun and all its planets and all their moons and the asteroids and some of the more local comets) would fit within a city block. When we look out at night, our eyes can detect the light from only 6,000 or so stars, the closest of which on this scale is an orange halfway across the country. The rest are too faint for us to see.

If we were somewhere else, though, the picture would be quite different. Just like people, stars tend to congregate in certain places. Most stars of the Milky Way Galaxy make a formation reminiscent of a UFO from a cheesy 1950s sci-fi movie—a thin disk with a big bulge in the middle. Peppered above and below this flying saucer are about 200 small blobs of stars. By small, I mean each blob is a scrawny little collection of 100 thousand to a million stars in a space only about 600 trillion miles wide. Tiny, aren't they? These so-called globular clusters are positively puny compared to the main flying saucer of the Galactic disk, which is about a thousand times wider and contains a million times as many stars.

The nighttime view from a planet around a star in a globular cluster would be pretty awesome if you were in just the right spot in just the right globular cluster. Perched high "above" the middle of our Milky Way's center bulge is just such a cluster. Aptly named Palomar 4, the fourth object catalogued in the 1950s by the Palomar Observatory Sky Survey, it is affectionately known by the astronomy in-crowd as Pal 4.

If you were lucky enough to live on a planet on the "bottom" side of Pal 4—the side that faces the rest of the Milky Way— you'd have the chance to see the giant sweeping hurricane of our spiral-shaped Galaxy take up easily half of your night sky. If there were some way to get to such a location in anything less than hundreds of thousands of years, you can bet that flights to Pal 4 would be overbooked for all eternity. For now, humans will have to content themselves with artists' conceptions of the view.

The vista from deeper inside a globular cluster wouldn't be quite so epic, but it would still make our Earthly night sky look pretty bleak. In these galactic minicities, stars are a bit more crowded than they are here. Here "a bit" means about "a thousand times." (This sort of language is why people often refuse to lend money to astronomers. "Can I borrow a dollar? Maybe a bit more?") With this many more oranges crammed into the

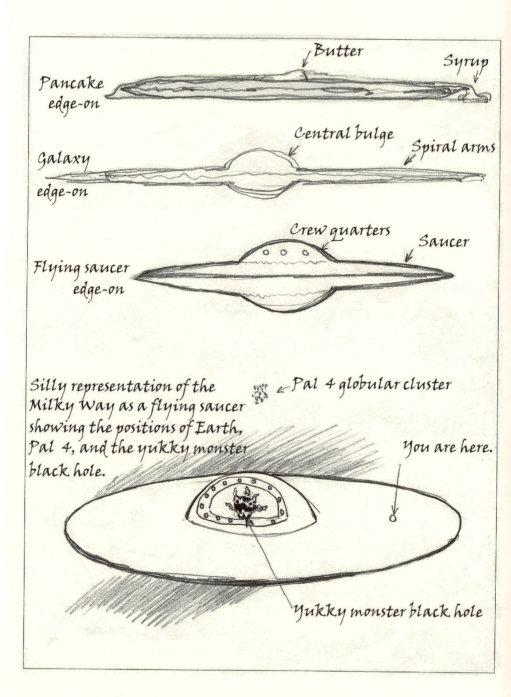

Butter

Syrup

Pancake edge-on

Central bulge

Spiral arms

Galaxy edge-on

Crew quarters

Saucer

Flying saucer edge-on

Silly representation of the Milky Way as a flying saucer showing the positions of Earth, Pal 4, and the yukky monster black hole.

← Pal 4 globular cluster

You are here.

Yukky monster black hole

Houston–San Francisco space, the night sky would perpetually shine with the light of a full Moon. Except, instead of being concentrated in a single silvery circle, the light would be spread out over a stunning array of bright yellow and red dots.

But if you're looking for a bright night sky, the kind that makes the lighting at a car dealer's lot look dim, you need to head to the center of the "flying saucer"—smack in the middle of the Milky Way. The trip there might take a while, though. The densest stellar population in the Milky Way Galaxy is so far away from Earth that light racing along at 670 million miles per hour would still take 25,000 years to get from there to here. By contrast, the Sun's light (also traveling at 670 million miles per hour) takes a little over 8 minutes to reach Earth.

At our Galactic core, the stars are up to a million times more crowded than they are in our neighborhood (or, alternatively, "a bit" more crowded than at the center of a globular cluster). That means that there would be a *million* oranges in a ball whose radius is the distance from Houston to San Francisco. Furthermore, the nearest one would be only about 20 miles away, and, as a result, it would appear thousands of times brighter than the one that is currently closest to the Sun. Our nearest neighbor, incidentally, is Proxima Centauri, and the light we see from it has traveled 26 trillion miles just to say "Hi."

Return the favor.

The night sky for a planet around one of those more central stars *would* be blazingly bright—nearly as bright as the day sky. Tragically, the Galactic center isn't all that hospitable a place. For one, there's a ravenous black hole with millions of times the mass of our Sun smack in the middle that is busily ripping apart everything it can get hold of. All this violent destruction lets loose a constant flood of high-energy (read: fatal) x-rays and gamma rays in its general neighborhood. As if that weren't enough reason not to relocate there, all your nearby stellar neighbors would be pumping out so much ultraviolet light, and occasionally exploding, that your chances of survival would be pretty small even if the black hole weren't there to vaporize you. And if *that* weren't bad enough, the collective jostling of all those crowded stars would very likely elbow your little planet right out of its comfortable orbit around its sun before you could say "Supermassive black holes and supernova shock waves, oh my!" And with no sun to orbit, there's no life.

All in all, the trade-off for a more impressive night sky is looking pretty unfavorable.

But take heart. We *will* ultimately get a different view of the night sky from here, even if not the insanely bright day-for-night conditions at the Galactic core. Like the other stars in the Milky Way, the Sun is orbiting the Galactic center, just as Earth and the other planets are orbiting the Sun. At its distance, the Sun makes a complete trip around the track in just over 200 million years. Earth, meanwhile, makes a single trip around the Sun in a year. But the Sun also has an added "upward" motion that will take it slightly higher than the flat flying saucer part of the Milky Way. Not "perched atop the Galaxy with a cosmic bird's-eye view" high like Pal 4, but enough to get us out of the Galactic smog that we're trapped in. In 15 million years (give or take), Earthlings—whatever they might be—will actually get a pretty stunning view of the bright, crowded center of our Galaxy. During the right time of year, the enormous central bulge of the Milky Way will rise like a luminous cloud and occupy a reasonably huge patch of our sky.

"How huge is it?" you ask.

It's so huge that it would take up about as much of your field of view as a VW Beetle would if you were standing 25 feet away from it. In other words, it'd be really, really obvious. Moreover, the rest of the Galaxy's "pancake" will be much less obscured as it stretches across the sky on either side of the huge glowing ball. Not the spectacular rising-Galaxy view from Pal 4, but pretty cool nonetheless.

Naturally, everyone on Earth will be so accustomed to such a view that they won't think twice about just how cool it is, and, unfortunately, nobody will be around to say, "You don't know how good you've got it, sonny. Back in *my* day, Earth was so immersed in the plane of the Galaxy that all these stars were blocked by dust."

Even if someone did, all the adolescent Earthlings would just roll their eyes.

If they did happen to roll their eyes upward, though, they just might notice that the night sky is still much darker than the day sky, and it might even occur to one of them that the Universe has billions of galaxies with billions of stars each, and all that light should add up.

Shouldn't it?

1.4 Why Is Space Dark? Answer #2: Because

Once upon a time, the Universe was endless. It had no birthday and no boundaries. Not that this is the way it really was, but it's how most people thought of it. After all, the idea that one day there was no Universe, and the next day there was, is kind of tough to wrap your head around. As for an edge, it just begs the question of what's on the other side, and whether the chicken wants to get there.

In a Universe like this, the darkness of night begins to pose a rather major problem. If it has been around forever, and if it goes on endlessly in every direction, then no matter which direction you look, your gaze should fall on a star. Stars are just suns, only much, much farther away (remember the orange in Houston and the orange in San Francisco). The common analogy is that if you were to stand in the middle of an infinite forest and look around, all your lines of sight would end on a tree trunk. Even if you don't buy into the idea that stars are like the Sun, you still have the problem of infinite points of light taking up your entire sky. The

night sky by this reckoning would be, according to some calculations, 90,000 times as bright as the Sun. For that matter, so
would the day sky.

And, once again, we'd be toast.

Clearly, though, we're very much alive, but we're still faced
with darkness at night, so this tells us something about either the
size of the Universe or its birthday or *something*. Great minds
throughout the ages have wrestled with this question with lim-
ited success. Really great minds. Astronomers like Kepler and
Galileo and Newton and Halley, and philosophers like Descartes
and Kant put quite a bit of effort into understanding what has
come to be known, strangely enough, as Olbers' Paradox. A paradox is basically just a logical contradiction, the sort of thing
that makes old sci-fi movie computers smoke and stutter. If, for
instance, you read a statement that said, "This sentence is false,"
this would present a paradox. Tell that to your computer next
time it's acting up and watch it self-destruct in a puff of illogic.

Similarly illogically, the paradox of the dark night sky is
always associated with a guy who, in 1823, came along centuries
after the first thinkers who pondered the situation. To compound
the illogic in the name, it happens that Olbers did some crude
calculations and was pretty sure that there was no paradox. The
night sky is dark because stuff in space absorbs a wee bit of starlight. Voilà! Problem solved! So not only is it named after just one
in a long line of thinkers, the darkness of night wasn't even much
of a logical issue to the one it *is* named after.

It turns out, though, that this isn't a particularly satisfying
answer either. Everyone knows that if a lamp shines steadily on
a piece of cloth, eventually the cloth will heat up. In an infinitely
old, infinitely big Universe—the type that everyone had desperately clung to for virtually all of history—whatever is absorbing
all that starlight would eventually soak up so much energy that it
would begin shining like the Sun, too, and the night sky becomes
blazingly bright again.

Olbers hadn't really thought of that.

Bizarrely enough, it was Edgar Allan Poe—yes, *that* Edgar Allan Poe—who came up with essentially the right answer without all the computations: even though the Universe might be infinitely big and infinitely old, stars themselves don't live forever and light takes some time to get here. Admittedly he said it much more eloquently than that. He was, after all, Edgar Allan Poe. But Poe was an amateur scientist at best, so his poetic treatment of the problem didn't get too much attention from the "real" scientists, most of whom wouldn't have bothered to read his poem on the matter. Instead it was a guy named William Thomson who later became known as Lord Kelvin (which sounds strikingly unlike either William or Thomson) who really provided the most rigorous proof for darkness.

What Poe intuitively grasped and Kelvin demonstrated mathematically was that what we see from the rest of the Universe has been traveling quite a while to get here. Sunlight takes over 8 minutes to get here. Light that we see from Pal 4 has been trucking across the Universe for 350,000 years. Light from the Andromeda Galaxy, our next nearest large neighbor, has been at it for over 2 million years. Kelvin himself had worked out a "liberal" estimate for the maximum possible age of a star: 100 million years. Turns out he was rather wildly wrong, but, hey, it wasn't a bad estimate for a guy who lived decades before we had any real clue what makes stars shine. Still, by his calculations, stars should pop in and out of existence rather often, and the Sun itself couldn't be more than 100 million years old. In an infinite Universe, a star that turned on today would remain invisible to us until its light managed to travel to our solar system. If that star were so far away that its light took more than 100 million years to get here (in other words, if that star were over 100 million light years away), we'd *never* see it because, by Kelvin's calculations, the Sun itself would have already died.

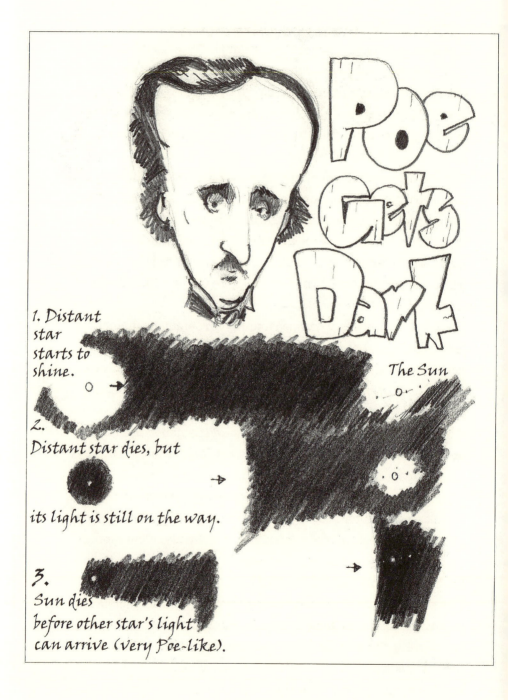

Poe Gets Dark

1. Distant star starts to shine.

The Sun

2. Distant star dies, but

its light is still on the way.

3. Sun dies before other star's light can arrive (very Poe-like).

Although the Universe itself might be infinitely big and old (an idea that Kelvin didn't much care for, anyway), what we can *see* of it is most assuredly not. At any given time, we just see a snapshot of the stars that have lived and whose light is reaching us at this point in history. Problem solved again.

Maybe.

In a classic Edgar Allan Poe story sort of twist, it turns out that the Universe is not infinite in age or size, but strangely it has no boundaries, and—even more strangely—the entire night sky really is blazingly bright. You just have to know how to look at it properly.

1.5 Why Is Space Dark? Answer #3: Actually It Isn't

Close your eyes. Breathe deeply. Now imagine the sound of one hand clapping. Imagine a tree falling in the forest when nobody is around to witness all the woodland critters panic. Imagine understanding the lyrics to your teenager's music.

Once you have achieved a state of total consciousness, you will be in good shape to understand the finite but unbounded, dark but bright Universe.

It all started with Einstein's general theory of relativity and then just got more bizarre from there. The Universe can't just be sitting there, Einstein's theory screamed out at him. It has to be doing *something*—getting bigger or smaller. Bothered by the idea that the Universe can't just *be*, he threw an extra number into his equations to make it behave.

Then astronomers noticed that the other galaxies in the Universe seem to hate the Milky Way with a vengeance. They're all running away from us, and the ones that are farther from us are going faster. So the Universe apparently *is* doing something after all. It seems to be getting bigger.

"Bigger into what?" your preschooler, spouse, neighbor, and everyone else will ask.

Breathe deeply and hearken back to that tree falling in the forest. This might take a while.

The easiest—and, therefore, terribly incomplete—way to think about what the Universe is doing is to imagine that you're a humble ant standing on a perfectly round balloon. Let's say that you're a pretty clever ant, and you decide to explore the balloon that you call home. You even leave a trail of your little ant pheromones to let you know the way back to your starting point in case you get bored with your trek. Antennae twitching with excitement, you head off. You never stray from your direction, and as far as you can tell, you're going in a completely straight line.

Now a really clever being—a small child—is watching all this, and she can plainly see that the ant isn't going in a straight line at all. The poor ant eventually walks full circle around the balloon, at which point its antennae quake with confusion. The twitching says, in ant language, "How the heck did I get back here? I've been traveling in a straight line!" The observant child giggles, knowing that the ant went in a circle.

A really sophisticated ant might realize that there's nothing to panic about and conclude that the object it's been walking on is actually curved. A few quick calculations, and the ant could even divine how big the balloon is and where its center is. Strangely, the center is in a dimension that the ant can't really take advantage of. It's just an ant, after all, and while it can cope with parts of the balloon that are in front of it and in back of it and to its sides, above and within are mysterious places that it can't explore. Sure, it can figure out that those dimensions *exist*, but it simply can't get there from here.

What's really cool is that the ant will never hit an edge of its balloon. It might explore the entire thing—the surface of a balloon has a definite size, after all—but it will never hit a wall beyond which it can't go. And if someone were to keep inflating the balloon, the ant would have more and more room to roam—possibly more than it could ever explore with its comparatively slow ant speed. Furthermore, things on the balloon would appear to be getting away from it. Even its own footprints dotted with ant-tracking scents would get farther and farther apart.

Again, a sophisticated ant would be able to figure out that the balloon is expanding, but since all it knows is a Universe of balloon material, it can't say what it's expanding *into*. Things just keep getting farther apart as the balloon itself stretches.

Now that you've digested this analogy, bump it up a dimension or two (or nine), and you've got our current understanding of the Universe, minus all the multidimensional mathematics. The problem is that we're really just ants. We don't grasp more than three dimensions, just as the ant can't grasp more than two. But we can tell that the Universe is doing something in those other dimensions, just as the ant could eventually determine its two-dimensional space was curved into a third and expanding. And, just like the ant, we could set out hunting for the edge of our Universe and never find it because, like the surface of a balloon, there's not a boundary.

But it's not an infinitely big balloon, either.

And this is the sound of one hand clapping.

So what does this all have to do with the darkness of night? Plenty. Our observable Universe contains an incredible amount of energy—unfathomable, but something that cosmologists can estimate nonetheless. All that energy was part of the package deal at its Universal birth nearly 14 billion years ago. As far as we can tell, there's not exactly a way for the energy to leak out, so it's been stuck inside the Universe all those years.

But the observable Universe is also expanding like the ant's balloon. Somehow our Universe is like an immensely elastic balloon that, when deflated, was really tiny. Tiny like the dot of this

"i" tiny. Tinier still. But it didn't seem tiny because that was the entire Universe. That much energy crammed into that small a space would make for a really impressively bright dot, and if you were hauled back nearly 14 billion years to see what the entire Universe looked like from inside, every patch of sky would be so insanely bright that staring directly at the Sun would actually be a nice break for your eyes.

Something made the Universe grow larger, though. Much larger. After its 380,000th birthday (give or take, but that doesn't make any sense at all because birthdays are marked by trips around the Sun, which didn't exist yet. But I digress . . .), the Universe was pretty big, so all that energy had a bit more room to spread out. Just as an aerosol can gets cooler when you release the pressure, the entire Universe cooled to a more temperate "face-of-the-Sun" kind of temperature by the time it was 380,000 years old. If you could be transported back to *this* time, every patch of sky would look like a patch of the Sun. This bigger, cooler Universe would actually look much darker than it did right after it was born. Interestingly, even though it appears darker, all that energy is still pretty much filling the Universe. It's just that the Universe is bigger, so the energy's more spread out—kind of like the ant footprints.

Now fast forward to today. All that energy's still out there, filling up space. But space is much, much bigger. ("Back in my day, sonny, space was smaller and everyone was much friendlier.") That energy is even *more* spread out.

Oh, it's still there. Make no mistake about that. Astronomers detected it over forty years ago, and it's everywhere they look. You just can't see it because your eyes can't pick up this particular type of light. If they could, the night sky and day sky would look pretty much the same everywhere you looked.

And children would never ask why it's dark at night.

Instead you'd be fielding questions about where babies come

from, questions that are far tougher than eleven-dimensional cosmology any day.

SMALL WONDER

Day and Night on Mercury

As magical as a trip to the Galactic center or a distant globular cluster would be, for the time being we humans will have to be content with the places we can get to: objects in the solar system.

One of the more fascinating day-night cycles can be seen from the planet that's currently hugging the Sun. Sure, it's incineratingly hot on the day side and so cold on the night side that Antarctica looks like a tropical paradise, but with the right shielding, a trip to Mercury is actually doable. The fun thing about Mercury is that its interaction with the Sun is similar to the case of the planets around Gliese 581. Similar, but not identical.

You see, tidal locking requires some pretty specialized dynamics, which is why the Sun hasn't managed to convince Mercury to perpetually point the same face to it. As it turns out, because its orbit isn't almost precisely circular, it's impossible for Mercury to be fully locked. The problem arises because things whip around a bit faster as they get closer to the thing they're orbiting—the Sun, in Mercury's case—and then slow down as they get farther away. Meanwhile, the object's rotation rate is relatively unperturbed, so sometimes it "over-rotates" and sometimes it "under-rotates" with respect to its orbit. In the case of our Moon, this gives Earth-bound viewers the impression that it disagrees with us as it slowly shakes its head back and forth every month, first showing a hint of one cheek, then the other. Meanwhile, astronauts watching Earth from the Moon would see it bounce ever so slightly up and down in the sky during the course of the month, an image that really makes me want to set up a huge holographic karaoke machine on the Moon. Just follow the bouncing planet . . .

Then there's Mercury. The Sun has been trying for over 4 billion years to get its nearest companion to show the same face to it, but Mercury's orbit is way off center and far from being a perfect circle. As a result, the Sun has to resort to the next best thing: forcing Mercury to show the same face to it every two Mercurian years. If you were born at high noon on Mercury, you'd be exactly two years old the next time it was high noon.

But the "over-rotating" and "under-rotating" problem would have fascinating consequences. If you just happened to be watching sunrise at the instant Mercury was farthest away from the Sun, you'd see the Sun rise—and it'd be a really bright Sun, too, appearing nearly 5 times brighter than it does from Earth. The Sun would reach higher and higher into your sky, growing larger and brighter the entire time, till it appeared nearly 11 times as bright as it does in Earth's sky. As you neared high noon, which occurs 44 Earth days into your Mercurian day, the Sun would appear to stop and back-track, almost as though it dropped something along the way. Then, having picked up whatever it dropped, it would start heading in the "right" direction again, growing smaller and fainter until sunset—88 Earth days after sunrise.

Only after a full trip around the Sun would you get to experience night, which is also 88 Earth days long. And when everything's aligned just right, you'd get to see a bluish-white dot in your night sky. Given that you'd be suffering from massive hypothermia, you'd probably be pretty interested in heading back home to that dot so you can experience a much shorter night that doesn't plunge to −300 degrees Fahrenheit.

SMALL WONDER

Keeping the Night Sky Dark

If you live in the city, you've probably noticed a distinct difference between your typical night sky and the night sky you see in the boondocks. On a clear night in the middle of nowhere, you can see thousands of stars, and the sky is so dark that you can see your shadow cast by planets like Venus and Jupiter. Smack in the middle of a big city, though, you'll be lucky to see anything other than the Moon and a few bright stars in the haze of the city lights.

This incredible difference is not just a passing concern for the relatively small group of people who like to stargaze. Bright artificial lights confuse the heck out of many animals, such as hundreds of thousands of baby sea turtles who wander the wrong direction in their attempt to head out to sea for the first time. Migrations and breeding cycles for dozens of species get thrown out of sync. The darkness of night is something that life—yes, apparently even human life—on this planet depends on. Artificial lighting disrupts our circadian rhythms, knocking our hormonal cycles out of whack, and the sometimes ambiguous day-night cycle even has been linked to increased cancer rates.

An entire organization—the International Dark-Sky Association (http://darksky.org)—is devoted to trying to recapture that darkness, not just for astronomers, but for everyone. Light pollution is just wasted money, anyway. Why light up the whole sky when efficient shielded fixtures actually do a better job of sending the light to the streets and parking lots we want illuminated? But it's more than just money. It's our humanity. We're children of that vast, awesome Universe, but we can barely see our cosmic ancestors through the false security blanket of artificial lights.

CHAPTER 2

Light

In the right light, at the right time,
everything is extraordinary.
—Aaron Rose

Now that you've sufficiently tackled the problem of night's darkness to everyone's satisfaction, the next question that your preschooler will throw at you is, "What's light?"

So you point at the lamp you've been using for your demonstrations. "That's a light."

"No-o-o-o," the child protests with a whine. "I know what *a* light is. What's the bright stuff coming off it? What makes it? How do we get all the different colors?"

The questions continue as your eyes glaze over. Somewhere in the depths of your mind, you know you used to wonder the same things as a child, and you wonder if your parents felt the same sense of hopelessness. Lining up your M&Ms by color, watching rainbows after thunderstorms, and seeing sunlight sparkle through a crystal figurine all made you curious about light. Unfortunately, all that innocent wonder was elbowed aside by More Important Things somewhere along the way, and you never got any answers except, perhaps, that light was something that moved really, really fast.

Whatever you do, don't reach for a dictionary at this point. The official definition for light is not, shall we say, enlightening. It's better to declare yourself an honorary astronomer and do some light reading instead, because, after all, "light reading" is basically what astronomy is.

You see, astronomers read light the way musicians read notes on a page. To a trained guitarist, those complex-looking patterns of lines and dots and numbers on a page represent music, whereas to someone like me, guitar tablature looks like the byproduct of hyperactive ink-covered grasshoppers. I could stare at it all day long and never figure out what it means because I've never been trained to read that code. But in the hands of a skilled musician, those strange markings become a musical story. Conversely, any musical story can be translated into a page of strange markings. Those markings can then be read and replayed by another musician—possibly thousands of miles away, or even hundreds of years later—who knows the code, and so the message crosses time and space.

That's essentially what light is to astronomers. We don't ever have to touch the objects we study to know what they are like. All we need to do is learn the language, and then we can decode the messages that the Universe has been sending out since its birth.

2.1 Codebreaking Basics

The first step in understanding light is, ironically, to close your eyes. Now think of something orange. Why do you suppose it looks orange? Imagine putting it in a sealed dark closet. Does it still appear orange?

The first thing that most adults think of when asked this question is a basketball, or maybe a shirt. Kids, on the other hand, will usually say, startlingly enough, an orange. In any case, it's usually something that won't retain its orange-ness when you turn out the lights. But if pressed to come up with more examples of orange things, people of all ages will usually say "fire." Put a fire in your closet and it's still very orange, and most likely incinerat-

MOM!
You CAN make orange in a closet!

ing all your other orange things. You can immediately conclude that the mechanism behind fire's orangeness is fundamentally different from that of the basketball. The basketball can't be orange unless there is already orange light present. The fire, on the other hand, is capable of producing its own orange light.

But fire is tough to control, so we need a less dangerous example of something that can create its own orange light. Something that fits the bill perfectly is a standard incandescent light bulb on a dimmer switch. If you don't have one handy in your house, call an electrician and get a dimmer switch wired in. It's just that important.

Turned low, the filament gives off a deep, dim orange glow. Play around with the dimmer switch and you can change the color from orange to a sort of yellowish, dusty salmon-y color to blazing white. Place your hand on the bulb (or, better yet, have someone else do it) as you brighten it and you'll immediately find out that white equals hot and orange equals (relatively) cool.

This simple experiment will lead you and anyone over the age of four to a very powerful conclusion about light. For light bulb filaments—and for things that behave like them—the color indicates the temperature of the object giving off the light. Fortunately, your shirts don't follow this sort of color-temperature relationship.

Congratulations! You have now completed lesson one in reading light. This is the astronomical equivalent of a music lesson where you find out that the piano keys at the left side of the keyboard give off lower pitches than the ones at the right side of the keyboard. In other words, it's a necessary start, but you won't be playing Beethoven sonatas anytime soon.

One of the things that complicates matters for beginning light readers is the fact that, just as not all musical instruments are pianos, not all light sources behave like light bulb filaments. A growing trend these days is toward compact fluorescent light bulbs (CFLs), bulbs that last longer and use less energy than their incandescent counterparts. If this trend continues, readers in fifty years will wonder how you can change a light bulb's color with a dimmer switch.

To look at a CFL, you'll note that the most striking difference is simply the shape. The light emitted, though, is basically white, very much like a standard light bulb at its maximum brightness. But anyone who has ever picked out clothes or vegetables underneath fluorescent lights can tell you that something's not quite the same. Out in the daylight or under standard bulbs, things just look ... well ... *different*. What's more, if you grab hold of a lit CFL, you won't require skin grafts later. To be sure, they get a little warm, but not nearly as toasty as a white-hot incandescent light bulb.

It turns out that if you play with a typical CFL on a dimmer switch, the color won't change. In fact, all it will do is turn on and off as you "brighten" and "dim." Even more annoying is that you'll nullify the warranty because all of that voltage variation wreaks havoc with a CFL's guts. You can't brighten and dim it unless you purchase a specific "dimmable" type of CFL, and no matter what kind of CFL you have, you certainly can't change its color.

Fortunately, there are plenty of examples of lights that operate without filaments. You can find scores of them at restaurants and bars, and they are universally (and usually incorrectly) called neon signs. Brilliant orange-reds, electric blues, and even yellowish dusty salmon-y colors scream out luminescent advertisements for beer. Put these signs (plugged in, that is) in a dark closet and they will still try to convince you in a variety of colors to buy a Guinness. But place your hand on tubes of different colors and you won't find any simple correlation between their color and their temperature the way you did for the incandescent light bulbs.

For the moment, lesson number two in understanding light is shaping up to be pretty frustrating. On one hand, you find that color and temperature are intimately related for one kind of

light-producer and, on the other, you find that they have no relationship at all in things like CFLs and neon signs. What's more
puzzling is that you can actually create the same yellowish dusty
salmon-y color with the appropriate "neon" sign (a helium light is
actually a pretty good one, but not so easy to find) and the right
tweaking of the dimmer switch!

So how can you tell which is which? And how on earth does
this relate to anything in the Universe other than light bulbs
and beer signs? To answer these questions, you need the cosmic
equivalent of the Little Orphan Annie Secret Decoder Pin.

But a simple CD or DVD will do just fine.

2.2 The Little Orphan Annie Secret Decoder CD

There's bound to be one in your house somewhere: a CD or DVD
that absolutely nobody can stand (who thought you'd enjoy
that for a birthday present?). Now's the time to put it to good
use. Your preschooler will likely be mesmerized for hours if you
simply tilt the CD back and forth underneath a lamp. Brilliant
rainbows will dance across its face and splash colors all over
the walls. Even your cats will dart into the room and attempt to
attack whatever is racing across the ceiling. As fun as all this is,
you need to stay focused on your quest to understand light.

If it's a sunny day, head to a dim room with some blinds. The
ideal setup is when the sunlight is trying to sneak in through
small gaps in the blinds. Hold your sophisticated scientific instrument (CD or DVD) between you and the tiny sunbeams and tilt it
in such a way that you can actually see a bright, narrow rainbow
radiating from the center to the edge.

The next step is a bit tougher. Somehow you need to get a similarly narrow beam from a normal, incandescent light bulb onto
your CD. This requires hiding under a dark cloth (uncomfortable)
or situating your CD in a box with a tiny opening in the end. In
short, you have to create something called a spectroscope, which,

despite the way its name sounds, is not a medical instrument. On the Internet there are tons of recipes for homemade spectroscopes using CDs—most of which call for a cereal box—and every last one of them says that it's important to get just a narrow beam of light bouncing off your CD.

If you don't want to spend the next half hour finishing off your kid's Froot Loops, you could just take my word for what you'll observe: the light from an incandescent light bulb passing through your painstakingly constructed spectroscope will be spread out into a rainbow.

Now before you complain that you have just gorged yourself on sugar and wasted a perfectly good Engelbert Humperdinck CD, a cereal box, some razor blades, and an hour of time, remember the basic light lesson from the previous section. Specifically, an orange light bulb is cooler than a white one. So you might as well use your newly constructed instrument to see if anything changes about the rainbow when you dim and brighten the light.

If you tweak the dimmer switch so that the light bulb is cool and dim, the rainbow you'll see in your spectroscope will be pretty much dominated by the reds and oranges, while the violets and blues will be essentially invisible. This probably won't be much of a surprise to you because, after all, the filament itself looks orange. As the light is brightened—and heated—the blues and violets (and all the other colors, for that matter, but most noticeably the blues and violets) get stronger.

Congratulations again! You have just found out two important things. The first is that light bulbs and the Sun both give you rainbows. The second is that the dominant colors on the rainbow depend on the temperature of the light bulb.

Now what about those other types of light? The neons and CFLs?

To get a good look at those, wait until night, that wonder of the Universe that you can now confidently explain because you read and acted out chapter 1. Once night falls, head out into the darkened world and find some outdoor lighting. Those dusty yellow-ish sodium lamps (low-pressure sodium, specifically) are great, as are the eerie blue security lamps that adorn many houses. It really helps if you are far enough away from the lamps that they just look like large dots because, once again, you need a narrow light beam for this to work right. Now hold your homemade spectroscope or just a plain CD near your face and tilt it to and fro until you see the "rainbow" from the lamp.

What you should notice pretty much immediately is that everyone on the sidewalk is staring at you, especially if you're holding a Froot Loop box up to your face. Simply explain that you are helping your kid with a scientific experiment (borrow a kid beforehand, if necessary) and hopefully that will buy you enough time before the police arrive.

As far as the light goes, though, you should find that you do not see a rainbow. At least not a full one. Instead, you'll see small dots of color. If you're staring at a sodium street lamp, you'll see very narrow bright smudges of yellow and orange and possibly a blue or purple dot. In between the colors will be blank spaces, completely unlike the unbroken rainbow you saw from your normal light bulb. A mercury vapor street lamp will display different colors but still in a broken pattern. A true neon sign (one that is a brilliant orange-red color) will give you an entirely different but still broken pattern of colors. And your household CFL will give you its own pattern.

It's important at this point to contain your excitement. There's nothing quite like someone on a street corner staring through a cereal box shouting, "Wow! Look at those colors! They're so cool!" to draw a crowd of concerned onlookers.

But it *is* impressive, because now you are able to identify the actual elemental fingerprints of the gases in those lights. No matter which low-pressure sodium lamp you look at, you'll see the same pattern. No matter which neon (true neon, that is) light you look at, you'll see the same pattern, and that pattern will be

"No, sweetie. Don't bother the interesting lady,"

different from sodium's. Mercury vapor lamps reveal their own light patterns. What this means is that if you are ever confronted in a dark alleyway by a mugger who demands that you tell him the composition of a distant street light, all you need to do is remain calm and politely ask for a CD.

As it turns out, every element under the right conditions will give itself away if you know how to decode its light. Those conditions are pretty specific, though. If you cram too much of the element together into something dense—like a light bulb filament— you lose the information about what it's made of. So it's got to be a low-pressure gas, and it's got to be energized somehow. But once those two conditions have been met, the atoms in the gas have no choice but to betray their own identity.

See? You're not the only slave to the laws of physics.

If you do have too much stuff crammed together, you're still in luck. Those light sources won't tell you their composition, but they will tell you how hot they are. So next time you see a complete, unbroken rainbow, you can divine the temperature of the light source.

This is not just a trick that can be used on light bulbs and restaurant signs. The night sky is filled with things that give off their own light—stars, nebulae, galaxies. If you're lucky enough to live in extreme northern or southern climes, you can even catch the aurora. Although your CD/cereal box spectroscope won't reveal all that many details about these light sources, more sophisticated (and correspondingly more expensive) versions can.

You see, simply looking at something in space is usually pretty pointless. Once Galileo pointed his telescope at the sky 400 years ago, the dots in the night sky were, for the most part, still just dots. The only difference was that he could see tons more of them in the same patch of sky. A view through the strongest telescope on the face of Earth isn't much better. The dot of the North Star— Polaris—still looks like a dot.

How unenlightening.

But if an astronomer splits the light from that dot of a star or some smudge of a nebula into its spectrum, she can tell whether she's looking at a hot, dense object, or a cloud of energized hydrogen gas, or even a hot, dense object that is surrounded by a cooler gas. Even more impressive is that astronomers can determine fairly precise temperatures and recipes for most of the things in the sky. We never have to stick a thermometer into Polaris to find out its temperature, and we never have to fly into deep space to retrieve a bagful of the Orion nebula to tell what it's made of. This is quite fortunate because we're stuck here on this planet, and the objects outside our solar system are so remote that getting to them is, at this stage in our technological history, nothing more than a child's fantasy.

2.3 More Than Meets the Eye

There are objects for which your cereal box spectroscope would be pretty useless because the light coming from them is, believe it or not, invisible. Not just faint, but actually *invisible*, no matter how much you try to magnify it. This comes as a huge shock to many people, including noted scientists. In fact, it's likely that the family of musician–telescope builder–astronomer William Herschel is still looking for his jaw that dropped to the floor when he discovered rays from the Sun that couldn't be seen.

The year was 1800. Herschel had been in his lab playing with rainbows because nobody had yet told him that there were More Important Things to do with his life. In his case, it helped immensely that he was colossally wealthy, having earned the title—and paycheck—of "The King's Astronomer" for discovering the planet formerly known as George (in honor of George III), but later designated Uranus. I think we'll all agree that third-grade boys are much happier with its current name.

Anyway, when you're the King's Astronomer, you're permitted and even encouraged to spend hours playing with light sources just to see what happens. In Herschel's case, he used triangular columns of glass known as prisms, rather than CDs, to make his rainbows. Just for grins, he'd been sticking thermometers into each of the colors of the rainbow to figure out if perhaps the color was related to the amount of heat coming through at that color. Like any good scientist, Herschel set up a "control"—a piece of the experiment that he "knew" the answer to—for comparison. The control in this experiment was a thermometer outside the rainbow, at a spot past red where there was no light coming through.

And, like any good scientific breakthrough, it was the so-called known answer that was the most astonishing: the thermometer reading there was the highest of them all.

Herschel dubbed the unseen heat source "caloric rays," and then proceeded, still like any good scientist, to play with them some more. Although he couldn't actually see them, he could manipulate them and make measurements about how these mysterious rays behaved. It turns out they behaved almost exactly like regular visible light. They could be reflected, absorbed, transmitted, and even bent as they traveled through something like a lens. The only difference was that they were invisible.

Since then, we've given his discovery a name that more accurately describes what it is: infrared light. It's called infrared because "infra" is the Latin term for "below," and "below-red" light just doesn't have the same scholarly ring to it.

Poor Herschel. None of the names he picked for things stuck.

He also never got to play with infrared cameras or night-vision goggles or any of the fun gadgets we have at our disposal these days. If he had, he no doubt would have wasted countless hours musing over the infrared images of simple things like animals and plants.

You see, just as the color of a hot light bulb tells you about its temperature, the infrared "color" of a somewhat cooler object can also tell you about its temperature. By the way, infrared images are all color-coded for our convenience because, obviously, we

don't really see infrared light. Typically an infrared image will have purple and blue for cooler regions and orange and yellow for hotter regions, which is really frustrating because we now know that hot objects emitting blue light are hotter than orange and yellow objects.

Perhaps infrared astronomers just like to be contrary.

Think about a zebra standing in full sun. The black stripes will absorb lots of sunlight and heat up, but the white stripes won't. The infrared photo of a sunning zebra will actually still have light and dark stripes, but they'll be reversed. The black (hot) stripes show up brighter than the white (cool) stripes in the infrared photos. And a black cat who's been lying on his side in a sunbeam will appear half yellow (hot) and half purple (cool) in infrared, but black all over visibly.

The upshot is this: if you ever get the chance to play with an infrared camera, go for it. And be sure to use an ice cube to paint "Hi Mom" on your forehead, or put a handlebar mustache and giant eyebrows on your daughter. Run a blow dryer through your hair, eat ice cream with your mouth wide open . . . do whatever you can to create a wide range of temperatures. The effect is pretty stunning.

Infrared light isn't all fun and games, though. For places in the Universe that don't reveal much about themselves visibly, infrared light gives astronomers incredible amounts of information. In fact, infrared light can get through some things that visible light can't, like the dust that obscures our view of the center of our Galaxy. It would also be impossible to get a peek at stars in the process of being born without a sneaky way to see through their dusty, cloudy swaddling blankets.

On the down side, there are things that allow visible light through that act like a brick wall to infrared light. Earth's atmosphere is one of them, blocking much of the infrared from the rest of the Universe. Even worse, the atmosphere itself gives off so much infrared at night that you couldn't see anything in space even if its infrared light *were* getting through.

Darned life-sustaining air.

But infrared astronomers are not to be outdone by something as trivial as dozens of miles of warm air enshrouding our entire planet. Instead, they use mountaintop observatories or even space telescopes to get much of their data. The images they produce are bizarre looking to us slaves of visible light. A region of sky that appears to contain a few innocent stars and a whole lot of nothing in between becomes an enormous blaze of complex filaments and swirls where cooler gases reside and interact.

If you're a bit disappointed at missing out on this view, you'll be delighted to know that most infrared observatories are more than happy to showcase their most photogenic findings on their websites. One of the best as of this writing is the cyber home of the Spitzer Space Telescope. The English version (as opposed to the version in the more technical language of astrophysics) can be found at http://coolcosmos.ipac.caltech.edu/. A gander at the images of galaxies, planets, nebulae, and animals (yes, animals—possibly even a gander) will likely fuel your search for your own infrared camera.

And, yes, eBay has plenty of listings for those.

Imagine what life would be like if our eyes could see infrared! Before you get too excited about the prospect, there are some issues you'd have to deal with. Sure, it would be great to witness the temperature map of a sunning zebra or be able to track a person by his warm footprints, but a pot of boiling water (and pretty much any hot meal or cozy campfire) would be painful to look at. Even the "dark" night sky would perpetually be a fairly bright haze through which you couldn't see the stars. Our ability to see fine detail would be lost, and the print on this page would be utterly pointless because the whole page would be a uniformly bland room-temperature sheet.

Naturally, if there's invisible light on one end of the rainbow, it would make sense that there's invisible light on the other side, too. Something that's above the violet, so to speak. Or, if you want to use Latin again, something ultraviolet.

Herschel didn't ever find this type of light, and it's not clear that he even looked. Most likely he got distracted by other shiny objects in his lab, so the glory of discovering ultraviolet goes instead to Johann Ritter in 1801. Ultraviolet was a bit more of a challenge to find. The thermometer trick that worked for infrared light only indicated cooler temperatures as scientists scaled the rainbow backwards from red through violet. What Ritter noticed was that a compound called silver chloride—something later used in photography because of its reaction to light—would darken when placed in the light. Blue light darkened it more efficiently than red; purple better than blue. And, you guessed it, something in the invisible portion just past purple worked best of all.

Just like that, all the textbooks on light and optics needed another major revision in less than two years.

It's really a shame that ultraviolet light has such a bad reputation. It's not fundamentally different from regular, visible light. It's not some toxic ray or nuclear fallout. It just happens to pack more of an energetic punch than either visible light or infrared, and that added energy just happens to be rather destructive to our skin cells. Fortunately for our skin, but *un*fortunately for anyone who wants to study the rest of the Universe in ultraviolet light, our atmosphere does a great job of stopping most of the ultraviolet light that comes at us.

Most—but not all. That's bad news for our skin, but good news for the few creatures that can actually see ultraviolet light. Birds and bees are two of them. (This is a fantastic escape from the usual "birds and bees" talk that you might have to give at some point. Just start talking about how the ultraviolet world looks to these creatures, and your audience will soon be so captivated that they forget what the original subject was.) Their view of the world is strikingly different from ours, with landing-pad patterns emerging on flowers that the human eye sees as solid yellow. Even some people are able to see a combination of visible and ultraviolet light, but this Superman-type of vision usually comes about with the surgical removal of a lens. While I'd love to get my own personal glimpse of the ultraviolet world, I think I'll have to pass on the major eye surgery.

But all of this is what astronomers would call "near ultra-violet" light, the type that's huddled on the rainbow right next to the normal violet that most people can see. Stray too far from that end of the rainbow and no Earthly creature can pick it up because none of it reaches the ground. The view in "far ultra-violet" light is distinctly dark down here under our cozy atmo-spheric blanket.

Once again, astronomers have to get above all that pesky air to get an ultraviolet view of the Universe.

It's worth it, though. Just as infrared light gives us a tempera-ture map of cool things (by "cool" I mean the temperature of hot lava or less—astronomers are masters of understatement), and the visible color of a light bulb or star tells us the temperature of hotter things, ultraviolet light can reveal what's going on in some of the hottest places in the Universe. A simple ultraviolet picture of our local, life-giving Sun shows astronomers a place vastly dif-ferent from the relatively quiet disk that you might see through welder's glasses or several layers of Mylar (the shiny silvery balloon material). Sure, you might see some sunspots, but those gradually changing blobs are nothing like what you'd see if you could catch the ultraviolet Sun in action.

Meanwhile, a telescope named SOHO (the Solar and Helio-spheric Observatory) has been in space since 1995 doing what you can't: busily staring at the Sun in both visible and ultraviolet light. Its main website (http://sohowww.nascom.nasa.gov) has several spectacular movies showing energetic blasts from the Sun, comets crashing into the Sun, and generally unbelievable solar activity. When you look at how puny Earth is compared to the hot, violent goings on in the Sun, you'll be pretty happy that you *can't* see it in ultraviolet. It might be enough to make you hide under a rock the rest of your life.

What'll really get you, though, is the x-ray view of our Sun. X-rays are not, in fact, something out of a comic book, despite a name that probably makes you wish you had a superhero alter ego. In a field that managed to come up with respectable, Latin-based names for infrared and ultraviolet light, why did we choose to keep x-rays?

Seriously, now . . .

X-rays are nothing more than a type of light that carries even more energy than ultraviolet, so much so that it isn't stopped by your skin. We ingenious critters have managed to make use of this fact to create x-ray machines to see inside the body without actually cutting into it. As everyone who has ever had an x-ray has learned, this type of light is pretty nasty in large doses. That's why the x-ray technicians cover you with a lead sheet and then run off and hide in a distant room while they check your arm for fractures.

Naturally occurring x-rays tend to come from the hottest, most violent places in the Universe, places whose temperatures are measured in the millions of degrees. Our own Sun apparently sprays out super-hot blobs of material the size of North America hundreds of times a day, activity that was totally unknown to us till an x-ray telescope named Hinode caught it in the act in 2006.

There's still no need to hide under a rock, though. X-rays don't make it through our atmosphere either. Come to think of it, the only types of light that do manage to get through Earth's atmosphere with any sort of ease are visible light and something called radio waves, which, despite their name, are not things you can hear. Instead they're a type of light at the other end of the energy spectrum, below infrared, and even below a type of light called microwaves, which are not places to reheat your food. Emitted naturally by the coldest places in the Universe, radio

Portraits of the Sun

Happy
Visible
Sun

CREepy
ULtRaViolEt
Sun

I AM BECOME DEA... uhh...
X-ray Sun

waves can get through just about anything—dust, gas, billions of trillions of miles of intergalactic space—to tell astronomers about the conditions where they originated.

The great thing about being a human is that since that first humble discovery of invisible rays in 1800, we've managed to figure out a way to create "eyes" to detect every possible type of light: radio waves, microwaves, infrared light, visible light, ultraviolet light, x-rays, and an even more energetic type of light called gamma rays. All seven of these make up something called the electromagnetic spectrum, which sounds for all the world like a battery of medical tests. ("Sorry, boss, I've got an appointment for a complete electromagnetic spectrum. I'll be out all day.") As an added bonus, our ability to manipulate these types of light has resulted in everyday gadgets like remote controls, night-vision goggles, cell phones, x-ray machines, radios, microwave ovens, and other things that cause us to mix up what the invention does with the type of light it uses.

Astonishingly, the amount of light that we can see without some kind of artificial aid is positively minuscule. If you represented the entire span of light from the lowest energy radio waves all the way to the highest energy gamma rays on a 2,000-mile-long scale, the window that represents what we can see is about an inch wide, and it's smack in the middle.

It's not that we've been short-changed, mind you. Remember that only visible light and radio waves have any success at getting through our atmosphere, anyway, so there's little point in having eyes that see the other types of light. Anyway, if we had eyes that could see radio waves, just the very nature of light would require each of our eyes to be several feet wide and separated by an even bigger distance in order to "focus" on anything. That might be a bit uncomfortable. Conveniently, though, most of the light pumped out by the Sun and many other stars lies within that tiny inch-out-of-2,000-mile-scale of visible light, so we really haven't missed out on much. Still, those other types of light have

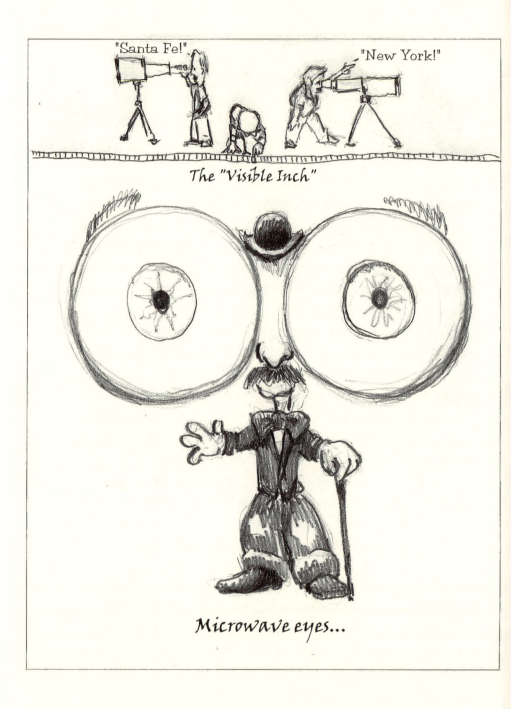

been handy in helping us understand the Universe we inhabit, allowing us to peer into places we were otherwise forbidden to see.

2.4 Evading the Question

"Okay, so now I know that there are lots of types of light and they all give us different information about the Universe," you complain. "But I've noticed you haven't actually explained what light *is*."

Darn. You weren't supposed to catch that.

Want the definition? Okay. Officially light is "radiation that consists of photons and travels through vacuous space at the speed of light and propagates by the interplay of oscillating electric and magnetic fields. This radiation has a wavelength and a frequency and transports energy."

Satisfied?

I didn't think so. Light in and of itself is a pretty tough concept to grasp. In the 1600s, Isaac Newton wrestled with it. He said light consisted of tiny particles racing through space. Around the same time, another gifted scientist and mathematician— Christian Huygens—said light was more like some kind of wave. Normal waves, like those in a pond, can do things that particles can't do, like bend around obstacles. Throw some rocks at a tree, and they won't land behind the tree. On the other hand, watch waves in a pond as they encounter a protruding piece of driftwood and you'll notice that they merge again on the other side. Another thing waves can do that particles can't is add up or cancel out depending on how the "high" points and "low" points match up.

An experiment with light in the early 1800s seemed to prove light was nothing more than a type of wave, and different colors simply correspond to different wavelengths, or distances between successive "high" points. So red light has longer wavelengths than purple, and infrared has even longer wavelengths than red.

The fact that light behaves like a wave is extremely useful. For instance, just about everyone can make the familiar "Eeeeeeeeeeeeeeeyowwwwwwwwwwwwwwwww" sound to mimic a race car as it speeds past. And the more musical-minded have mentioned how out-of-tune a police car or ambulance seems to become when it passes by. The change in pitch is a consequence of the wave nature of sound. Higher pitch sounds are just shorter wavelengths where the waves are bunched up closer together. Lower pitch sounds are nothing more than longer wavelengths.

If an ambulance is moving toward you, the sound waves from its siren essentially get bunched up in front of it, giving the siren a higher pitch than normal. When it passes by and is moving away from you, its sound waves are essentially getting stretched out behind it, so you hear a lower pitch.

In the 1840s, a guy named Doppler figured out that you can actually calculate the exact speed of an object if you can figure out two things: (1) What note you're supposed to hear and (2) what note you're actually hearing. The exact same trick works with light waves, except instead of notes, you need to figure out the exact colors.

More exact than just whether it's red or blue.

I mean *exact* colors.

Remember those CFLs and sodium-vapor streetlights that you played with earlier? The wavelengths for the bright-colored dots that you see have been measured to unfathomable precision. You might see, for instance, a bright orange-ish yellow dot. Someone in an introductory astronomy class would be taught to assign a wavelength to that color in a unit of measurement that is completely unfamiliar to most mortals: 5900 Angstroms.

Yes. Angstroms. For most students, it starts with angst and just keeps going from there. An Angstrom is an unimaginably tiny length. The humble inch can be sliced into more than 250 million Angstroms. Needless to say, you don't whip out a ruler to measure wavelengths of visible light.

The human eye can tell the difference between yellow (about 5900 Angstroms) and green (about 5500 Angstroms) light, but you can't visibly tell whether the yellow light is 5850 or 5900 Angstroms. To get the exact wavelength requires professional light-measurers. These scientists can tell whether something is 5900.000 Angstroms or 5900.001 Angstroms.

They're just that good.

Anyway, back to your cereal-box spectroscope. What you see as a bright orange-ish yellow dot from a sodium-vapor street lamp actually turns out to be two dots with the incredibly precisely measured and catalogued wavelengths of 5889.973 Angstroms and 5895.940 Angstroms. But if that streetlight begins to move toward you (let's hope it won't), you'd measure a shorter, smaller wavelength, something that leans slightly toward the greens and blues.

The technical term for this "higher-pitch" that you'd measure is blueshifted. And, likewise, if the streetlight is moving away from you (which is still creepy, but at least it's not coming after you anymore), its light will all appear at wavelengths that are longer than the catalogued values, or redshifted. Just as with sound waves, if you know what the wavelength's supposed to be (the catalogue value) and you measure the exact wavelength that you see, you can figure out how fast the streetlight's moving.

This is essentially how the cops catch you when you're speeding, by the way. Their "radar gun" knows the exact wavelength of the signal when it leaves the gun. Then it bounces off your moving car and heads back to the officer with a slightly different wavelength. The radar gun's computer does some quick calculations and the next thing you know, you're explaining to a uniformed stranger just how badly you need to find a bathroom.

One thing you should never do, though, is attempt to use this

newfound information to get out of a citation for running a red light. "Gee, officer, that red light must have been so blueshifted that it appeared green to me." I'm sure the fine for going half the speed of light is far more than the fine for running a red light. On the other hand, at that speed, you probably wouldn't get caught.

There is one slight complication to using the Doppler shift with light. To figure out the speed of the moving object, you actually need to know how fast light travels. Galileo tried to figure this one out by flashing lamps back and forth across mountaintops and all he succeeded in doing was measuring his assistant's reaction time. His conclusion? Light moves really, really fast.

It wasn't until the time of the U.S. Civil War that James Clerk Maxwell (not from the United States) put some equations dealing with electricity and magnetism together that described lots of things about light, including its speed, which is about 186,000 miles per second. Like I said, it's really, really fast. Oddly, the collection of four equations that currently adorns the fronts of nerdy, but clever t-shirts proclaiming "And God said" [insert big pile of equations here] "and there was light," are called Maxwell's equations, even though Maxwell himself formulated only one of them and basically just collected the rest together in a coherent package.

Then again, the "Why is it dark at night?" problem is called Olbers' paradox.

And the planet George is now known as Uranus.

Sometimes science makes very little sense.

But back to astronomy. By 1870, astronomers and physicists were in absolute ecstasy because not only could they divine temperatures and compositions of things trillions of miles away, but they could also figure out how fast and in what direction those things were moving. All this after a very noted philosopher by the name of Auguste Comte declared with great certainty that astronomy "will leave us always at an immeasurable distance from understanding the universe."

I've long wondered what I would say to Auguste Comte if I could meet him. I think most civilized, well-mannered folks would agree that the appropriate response would be, "Neener neener neener." I might toss another "neener" in just for good measure.

But everything isn't as nice and neat as I've led you to believe. Later experiments showed that light was even more bizarre than ever imagined: it apparently doesn't need to travel in any kind of medium. This amazing result was found, naturally, during an experiment to measure the properties of the medium that light travels in, so, in that respect, the experiment was a total failure. Surely light had to travel along *something*. Water waves, for example, need water to get from one place to another. Sound waves need some kind of material to travel through. Waves on a Slinky need, well . . . a Slinky. But light waves seem to make their own medium as they go, and they're able to blithely get through the vacuum of space without water or Slinkies.

But wait! There's more! While people were still trying to digest the concept of a medium-less wave, Einstein came along in the 1900s and confounded everyone with a few experiments that seemed to prove that Newton was right after all: light also behaves like little pieces of stuff racing through space.

So it's a wave.

But it doesn't actually *wave* anything.

And it's a particle.

Furthermore, the particles with the shortest wavelengths carry the most energy.

Once again, this is the sound of one hand clapping.

And I'll bet you're still annoyed that I have yet to explain what light is, only more about how it behaves.

The good news is we don't really have to explain what light *is*. We just need to know what it does and how it interacts with other things and it will tell us nearly everything there is to know about the places that it's been.

I guess you could say that light is the cosmic gossip, and as you'll see, some of its truths have been stretched quite a bit since the birth of the Universe.

2.5 Making Light of the Universe

Remember the ant on the inflating balloon from chapter 1? We need it to return for a moment so we can see what light tells us about the origin of everything. Actually you can leave the ant behind. We just need its amazingly elastic balloon that can shrink to a point when deflated.

Also recall that all of the energy in the entire Universe was once shrunk into that point, at least if something called the Big Bang theory is correct. That would make for an intensely hot place, and, as you now know, that would mean the entire Universe would have to have been filled with the most unimaginably short wavelengths of light. The gamma rays that accompany a nuclear blast would be, by comparison, really long, harmless wavelengths of light. The waves in the Early Universe would be infinitesimally small, and they'd be all over this also infinitesimally small balloon.

Although you most likely can't get an infinitesimally small balloon, you can still see what will naturally happen to the light as the balloon begins inflating. Draw some small-wavelength squiggles all over a normal balloon. Then blow up the balloon a bit.

Notice what happens to the squiggles?

"I don't have to notice!" you declare gleefully. "I've already figured it out because I know that an expanding Universe gets cooler, and cooler things have longer wavelengths associated with them."

Fine. Be that way. But it's still fun to blow up a balloon and see the effect. The formerly short waves stretch out with the balloon and become longer waves. Blow the balloon up more and they become even longer.

At this point in history, our Universe is billions of trillions of miles across, and it is filled to the brim with long wavelengths associated with the low-low temperature of −455 degrees Fahrenheit. That's −270 Celsius to most residents of the globe, and 2.7 Kelvin to scientists, who just like to be different. This is down from the toasty 100 million trillion trillion degrees when it was a mere billionth of a trillionth of a trillionth of an inch wide.

Like I said, it's a really stretchy balloon. And apparently it can withstand some pretty crazy temperatures.

"Wait a minute," you protest. "I thought you said light didn't have to travel on anything, and now you've got the waves being stretched by this rubbery balloon material."

Did I say that? What I meant to say is that light essentially becomes part of the stretching fabric of the Universe as soon as it's made. But did you really want to have to contend with the concept of some elastic fabric of space-time along with the whole wave-particle thing?

The instant coupling of light to that balloon has made for some pivotal, but highly confusing, observations in the history of astronomy. When you turn on a lamp in your living room, the light will get to you in no time flat (that's the technical term for a hundred millionth of a second or so). In that time, our Universal balloon doesn't stretch much, so the wavelengths you see are not measurably different from what left your lamp. (Even so, Earth makes a slight dent in the balloon, keeping it from stretching as much as the rest of the balloon, but that's an entirely different story for an entirely different book.)

When a star in the outer parts of the Milky Way explodes, it takes a bit longer for the light of that event to get to us—tens of thousands of years, perhaps. But even in *that* time, the balloon doesn't stretch much, so the light that you see is basically the same wavelength as what left the star.

When we look at a distant galaxy, though, the light has been trucking across the Universe for quite a while, possibly even billions of years. The balloon's stretched quite a bit in that amount of time, so the light we see has much longer wavelengths than it did when it left that distant galaxy. We see a redshift, not because the galaxy is moving away from us (à la Doppler), but because the light has been stretched as it's been riding on the balloon.

If you look at an even more distant galaxy, its light has been working its way toward us for even longer and has stretched even more.

This observation—without the advantage of the spectacularly illustrative balloon demonstration—was made in the 1910s and 1920s by Vesto Slipher and Edwin Hubble for a handful of galaxies, and since then it's been made for millions of them. Closer galaxies have less-stretched light (or smaller redshifts), while the more distant ones whose light has had to travel along the inflating balloon-iverse for longer show more-stretched light (or bigger redshifts).

While most people know Hubble's name (he's had an entire space telescope named for him, after all), almost nobody has heard of Slipher. Apparently the presence of redshifts in the light coming from galaxies just didn't strike him as something worth pursuing. On the bright side, there's a lunar crater that bears his name. Personally, I think I'd pick the space telescope over a big hole. On the other hand, the lunar crater will still be there after the Hubble Space Telescope is reduced to shrapnel.

Anyway, in the early days after this observation, astronomers were highly puzzled because nobody had any idea that redshifts could arise from anything other than motion. So if all the galaxies in every direction had some kind of redshift, that seemed to imply the Milky Way was smack in the center of the Universe and all the other galaxies must have really hated us. Even more puzzling was that all the galaxies at one distance were apparently leaving at a certain speed, and all the ones that were farther away were leaving at an even faster speed. So not only did it look like we were at the center, but it also appeared that the rest of the Universe knew we were at the center and knew how far away we were.

It was really creepy.

The balloon thing—while still a bit weird to grasp ("Right . . . we're living in a stretching balloon of a Universe . . .")—at least explains those redshift observations without being quite so creepy. It also explains the observation of the long-wavelength light everywhere you look, something that would make both the day and night sky almost uniformly bright if only you could see what it looks like in microwaves, mostly microwaves with a wavelength of about 1/25 of an inch.

Imagine! The Universe balloon has stretched from an immeasurably hot and small dot to this vast, cold place billions of trillions of miles across. As the Universe has grown, those original tiny waves that filled it have stretched to all of 1/25 of an inch.

That initial light had a really, really short wavelength.

Really.

It was also a really, really energetic particle.

And despite all you now understand about what light does, you probably still don't understand any more about what light is. But at least now you can line up your M&Ms by the predominant wavelengths that they reflect.

 **SMALL WONDER**

Why Is the Sky Blue? And Why Are Sunsets Red?

Great questions. And identical. So let's kill two questions with one answer, shall we? Both the blue sky and red sunset are the result of the way our atmosphere interacts with light.

As you already know, if you look at white incandescent light bouncing off a CD, the different colors are separated. Different wavelengths of light also are emitted by different elements.

Different colors basically have their own distinct personalities, and not just in the ways we've looked at so far. They also can act differently when they encounter stuff like small particles of dust, which are abundant not only on my bookshelves, but also in the air you breathe and in the incredible vastness of interstellar space.

Because of its longer wavelength and more easy-going personality, red light tends to be bothered less by dust than more energetic, hyperactive blue light. Sure, it might get dimmed somewhat, but it gets through it reasonably well. Blue light, on the other hand, has a panic attack when it hits molecules in the air or even in space. Instead of passing through relatively unperturbed, it finds itself kicked into a different direction entirely.

So, imagine a clear, sunny day with a beautiful blue sky. The Sun is churning out the entire rainbow of colors (along with some things that don't get through our atmosphere), but poor blue light gets kicked around by those bully air molecules. Some of that blue light *might* have been heading through the atmosphere over your head, but part of it gets redirected down to you. Some of that blue light *might* have been heading through the atmosphere to your left, but part of it gets redirected to you. Everywhere you look, some blue light that was originally heading to a place *other than you* finds itself kicked toward you.

Now, about those red sunsets. When we look at a sunset, we're looking through a really thick slice of atmosphere, thicker than if we're looking at the sun straight overhead. That means even more blue light has been redirected, and the red light that isn't bothered as much by all those bully air molecules makes it to you just fine.

Tragically, we're so used to red sunsets and blue skies that we take them completely for granted. But on the Moon, there isn't an atmosphere to kick the blue light around. Instead, if

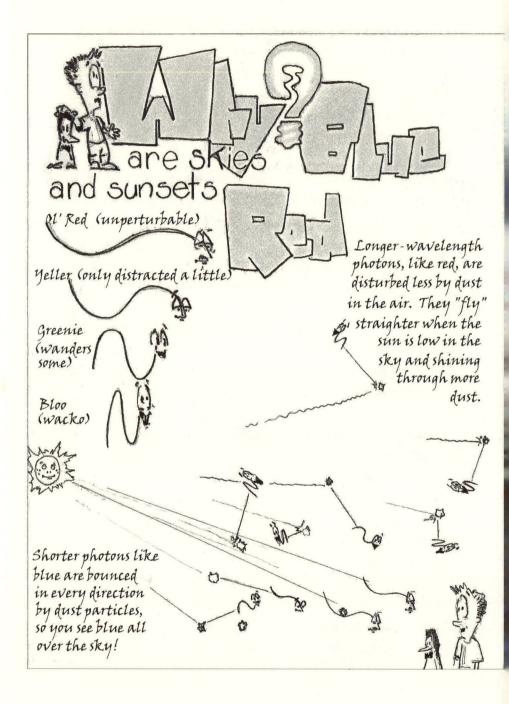

Why? are skies and sunsets Blue Red

Ol' Red (unperturbable)

Yeller (only distracted a little)

Greenie (wanders some)

Bloo (wacko)

Longer-wavelength photons, like red, are disturbed less by dust in the air. They "fly" straighter when the sun is low in the sky and shining through more dust.

Shorter photons like blue are bounced in every direction by dust particles, so you see blue all over the sky!

you're not staring at the blazing disk of the Sun, you're seeing the blackness of outer space. No blue sky. No deep red and orange sunsets. Just blinding light and inky darkness.

 **SMALL WONDER**

The Early Universe—A Made-for-TV Movie!

If you want to see the most epic show in all of history, find a television that actually has an antenna. This task is getting more and more difficult as we've completed the transition from analog to digital signals, but see if you can get hold of one anyway. Then disconnect the television from any cable or dish input and tune it to some random channel.

Now park yourself and your family and friends on the couch with some popcorn and enjoy "The Early Universe."

As it turns out, a large portion of so-called static or snow or noise is actually your television tuning into that low-energy, long-wavelength remnant of the tiny, hot Universe. Otherwise known as the Cosmic Microwave Background Radiation, it's at just the right wavelength to be picked up by your television antenna.

I know. It's not the most interesting plot line. But it is pretty mind boggling when you think that you're actually watching leftover light from the beginning of the Universe.

CHAPTER 3

Stuff

If you wish to make an apple pie from scratch,
you must first invent the Universe.
—Carl Sagan

A t some point in your life, you will walk into a room. Perhaps it's your room; perhaps it's the garage; perhaps it's your child's room. And you will survey the contents of the room and, in great frustration, you will cry out, "Where did all this *stuff* come from?!"

In the case of the child's room, the answer is probably her grandparents. But where did *they* get it? At least twelve different stores in the megamall, of course. But where did those stores get it? Probably from their distribution warehouse, which got it from the manufacturing facility, which got it from the various raw materials scraped from the ground, which got it from . . .

. . . where, exactly?

If you really want to trace the lineage of all the stuff in your house and in your food and even in your own body, you'd better be prepared for a long voyage. Much of it has been riding around the Universe for nearly 14 billion years; the rest was manufactured in great cosmic element factories over 5 billion years ago.

Even more puzzling than where all your stuff came from is this: nobody is really certain why stuff even exists at all.

3.1 It's Element-ary

First you take a small Styrofoam ball and paint it black. Then you take another small Styrofoam ball and paint it red. Then you glue those together. Then you take a wire coat hanger and make a loop, dangling the glued Styrofoam balls in the middle of the loop with some floss. Then you place a small yellow bead on the wire, and, voilà! You now have a fifth grader's homemade atom ready to hit the school science fair.

Except real atoms aren't anything like that. Oh, sure, they generally have three different types of particles—protons (red Styrofoam balls), neutrons (black Styrofoam balls), and electrons (yellow beads)—and the basic layout is at least crudely represented, but that's where the similarity ends.

The reality is, like pretty much everything else in the Universe that we take for granted, far more bizarre than you can imagine. The most glaring problem is that the relative dimensions are utterly wrong. If you want your fifth-grade niece to make a scale model of an atom, she might have difficulty fitting it into the school auditorium. Or even into your county. A tiny sixteenth-of-an-inch electron bead would have to be accompanied by a couple of 5-foot-wide neutron and proton balls to get the basic sizes of the particles right.

If that's not hard enough to haul into the school, now things get really ugly. To get the scaled distance right, the loop of wire holding the electron bead is going to have to be over 100 miles across.

It's not just the vast interstellar distances that are mind-boggling (remember the orange in Houston and the orange in San Francisco?). Even tiny subatomic distances are unbelievable when you try to blow them up to sizes we can understand. In fact, when you look at the Universe, most of it consists of a whole lot of nothing.

A SENSE of SCALE

You have a BIG atom.

If THIS is your electron...

...THIS is part of your proton,

and your atom is 100 MILES ACROSS!

Baltimore
Washington, D.C.
Alexandria

So, how much space is that? Enough to cover the 48 contiguous U.S. states...

...890 feet deep!

All except

ECONO-BOX

65 cubic feet of the space in that sphere would be TOTALLY EMPTY.

They certainly don't tell you *that* when you go to the store. "Consumer advisory: All products for sale contain 99.999999% empty space."

But what an amazing variety of things you can make with all that empty space and three types of particles! All the calcium atoms in your bones consist of 20 of those red balls (protons) and 20 yellow beads (electrons) and typically 20 black balls (neutrons). For calcium, and for any element, the number of protons tells you what the element is, so calcium is element number 20. The hydrogen that makes the H's in the H_2O that you drink has only 1 proton and 1 electron and usually no neutrons, so it's element number 1.

Element number 79 is the gold in the earrings you want. And the lead that medieval alchemists so desperately tried to change into gold is element number 82, which usually has 125 neutrons cloistered together with the protons in the central ball we call the atomic nucleus. If only they'd known how to yank 3 protons and 7 neutrons out of that central ball, they'd have made their fortunes. On the other hand, despite their superior understanding of how atoms behave, today's nuclear physicists aren't exactly churning out tons of gold either.

So where *did* it—and all the other elements, for that matter—come from?

3.2 Element Factories

If you randomly take 100 atoms out of your own body (don't worry—you're dropping about 4 billion trillion atoms each day), about 62 of those will be hydrogen. Take 100 atoms from the Sun, and you'll find that about 90 of them are hydrogen. Take 100 atoms from just about anywhere in the entire Universe and again you'll find that most of them will be hydrogen. Simple, everyday hydrogen, with one proton, one electron, and usually not a single neutron.

Almost everything else in the entire Universe consists of helium, but unless you're actively inhaling some helium to make a Mickey Mouse voice to entertain your concerned friends, your body has no helium. What you do have instead are things like oxygen, nitrogen, carbon, and dozens of other elements that you need a pronunciation guide to get through (molybdenum, for example). The rest of the Universe has all these elements as well, but just not quite in the concentration that you have them.

For the sake of simplicity, let's look first at something that *isn't* the simplest atom. Oh, sure, you might think the easiest elemental story to trace would be hydrogen's, but as is the case with so much else in life, appearances can be deceiving. So instead let's take a look at everything else.

To understand how to create something like helium, you first have to erase a few scientific "facts" from your mind. These are things you've held dear for probably your entire primary and secondary school career, but like that old homecoming corsage, it's time to let them go. So concentrate on the swinging pocket watch while repeating these statements:

Like charges repel.

Matter can neither be created nor destroyed.

Now, when I count to three and snap my fingers, these two facts will be erased from your mind. Ready? One . . . two . . . three . . . [snap!]

About those charges. Around the time you were getting (or giving) that homecoming corsage, you were probably making excuses for your date by insisting to your parents that opposites attract. Fair enough. That proton sitting in the nucleus of a hydrogen atom certainly holds on to its attendant electron. The proton has some property to it that we call a positive charge. The electron has some property to it that we call a negative charge.

Blame the ancient Greeks and Ben Franklin for this terminology. The word "electron" is actually the Greek word for amber, of all things, and it came to be associated with electricity (which comes from that same Greek word) because you can rub a piece of fur and amber together and give yourself a pretty good zap. If you don't have amber or fur sitting around your house, you can do the same thing by shuffling across the carpet on a dry winter day and then reaching for a metal doorknob.

If you're looking for a less painful way of exploring the wonders of electrical charges, you can rub a balloon on the head of someone with incredibly fine hair instead. Little kids love this, especially if there's a mirror around. Teenagers . . . not so much.

What happens is that some of the electrons from the hair jump to the balloon, leaving the hair positively charged and the balloon negatively charged. Remember, though, the assignment of "positive" and "negative" is totally arbitrary. We could have called them A and Z, or yellow and purple. But we picked positive and negative, as though they were attitudes.

(So a hydrogen atom runs into the police station and exclaims, "Help! Someone stole my electron!" The officer says, "Are you sure?" Hysterically the hydrogen atom screams, "Yes, I'm *positive*!" [rim shot])

Yes, well . . . now that you've gotten some good video of hair standing at attention, move the balloon from side to side and watch how all the hairs keep reaching toward the balloon. Add some music if you like. Now take the balloon away. All the like-charged hairs seem to really, really hate each other at this point, even without the balloon's negative charges to pull them balloon-ward. In fact, so great is their repulsion for each other that they'll defy the entire tug of Earth just to keep from touching any of the other hairs.

And thus a scientific "truth" that opposites attract and like charges repel is born and perpetuated throughout science classes everywhere.

Now I'm not denying that the negatively charged balloon and positively charged hair will draw each other together (you can actually see this happening). But it's such a superficial attrac-

Opposites attract and likes repel...

...until they don't...

Crush protons together
strongly enough to
nearly touch and...

...the "strong nuclear force"
will glue them together.

tion. That puny yellow bead of an electron is over 50 miles away from the 5-foot-wide proton that gives the element its identity. Do you honestly want to trust *it*? This scrawny thing that can be removed so easily from your child's hair with a balloon? Or are you looking for something more stable? Anyway, if like charges hate each other so much, how can 20 of them possibly snuggle so close to each other inside the nucleus of a calcium atom? And why doesn't the electron get closer to the proton if it likes it so much?

If it never occurred to you to ask your science teachers these questions, now's the time to make plans for that high school reunion.

First let's look at the protons. They act remarkably like small children sometimes. You try to convince them that they'll really like something (not broccoli, mind you, but something that they really *will* like—ice cream, for instance), and they resist it. Try harder to get them to take a bite, and they resist it even more. The harder you try to put the child and the treat together, the greater the resistance. If by some miracle you actually manage to sneak a bit of it into the child's mouth, you'll find the strong resistance suddenly turns to an even stronger attraction. Then you won't be able to separate the two.

Believe it or not, protons are just as childish. Try to push two protons together and they will fight tooth and nail (okay, so they don't have teeth or nails . . .) to stay apart. But once you get them close enough, they pull each other together with the strongest sticking force known to mankind. Even stronger than a small child's attraction to ice cream.

So what's this amazing sticking force called? Something intellectual-sounding and Latin? Not exactly. Since the force acts on particles that live in the nucleus, it's called a nuclear force. And since it's really, really strong, it's called . . . well . . . strong. Put it all together and you get the strong nuclear force, and it's better than the best adhesive around.

Interestingly, even though the two protons are stuck together with the strong nuclear force, they still have those like charges that are trying feebly to push each other away. Their repulsion is pretty pathetic by comparison, though. It's only about 1/100th as strong as their attraction, and it only gets to be a problem for a nucleus if you try to cram too many protons together. Then really fun things begin to happen.

But that's another story.

What about those electrons? Don't they want to be part of the fun in the nucleus as well? You'd think that their natural attraction to the protons would be enough to let them crash that little nuclear party, but they can't. You see, they're not exactly beads. They're more like wave-beads. Or bead-waves. Or wave-icles.

Did I mention that things on the subatomic scale are more bizarre than you ever imagined? At this level (and, truth be told, at every level, but it's most noticeable at this level), the distinction between a particle and a wave is not so easy to make. Just as light can behave like pieces of stuff or waves depending on how you want to measure it, so can these tiny electrons that we tend to represent with a small bead. Apparently nature is filled with split personalities.

The idea of an actual *thing* acting like a wave is hard to swallow, but electrons have been known to bend around obstacles and even add up or cancel out like regular waves. This makes life interesting for an electron if it's trying to run laps around a proton.

Think of an electron on a circular race track where the starting line and finish line are at the same location. Let's say the electron's wave begins at a low point—or trough—just at the starting line. If the track is exactly the same length as the electron's wavelength, the electron will wind up at the finish line when its own wave returns to a low point. But if the length of the track is exactly half the electron's wavelength, the electron will wind up at the finish line when its own wave is at a high point. This high point will actually cancel out the low point that it started with because, believe it or not, the electron is also still at the starting line. It turns out that unless the track is precisely long enough to fit in some whole number of electron waves, the poor electron will always wind up canceling itself out of existence.

Weird.

What this means is that wire hoop with the yellow bead can't be just any old length. It has to be just the right length to fit exactly 1 or 2 or maybe 17 or 455 electron waves on it. But it can't be shorter than 1 electron wavelength around or the electron will die.

At this point, you need to take a break from reading this and make some notes for your current or future children. There will be a time that you hear, "But Mo-o-om (or Da-a-a-ad), don't you know that opposites attract?" as your horror at their choice of dating material becomes apparent.

And your science-savvy response can be (calmly, of course), "Yes, honey, I know. Just look at the atom. The proton and electron do seem to have a pretty strong bond, don't they?"

At this point, your teenager will wholeheartedly agree. Just continue calmly, "But, on the other hand, you'll notice that the electron always keeps a safe distance. If it tries to get too close to the proton, it will *die*." Be sure to emphasize that last word. Then, soothingly remind the teen that even though protons have like charges (and probably similar interests and goals), they are stuck together like glue because some forces are just more powerful than others. "Better to find someone a bit more like you if you don't want to disobey the laws of physics, okay?"

No, it won't convince teenagers. But it is fun to be able to give them a quick lesson on atomic physics when they least expect it.

At this point you're probably thinking, 'Okay, so I've got some good information on annoying any future teenagers in my household, but I still am a bit fuzzy on how elements are made.'

Actually you've got most of it already. Remember that the

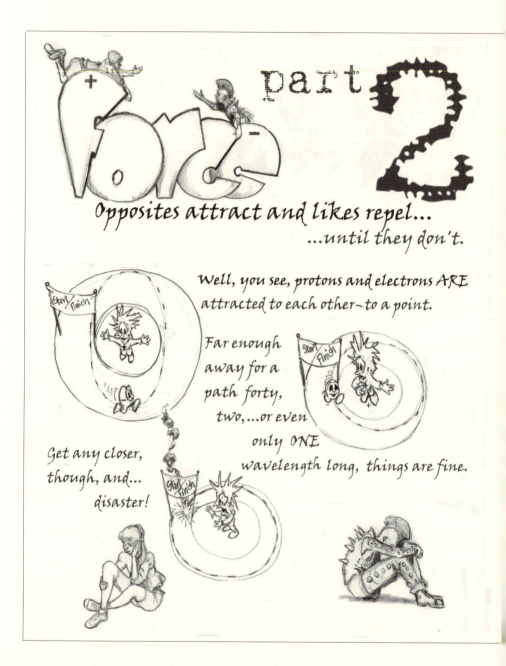

Force part 2

Opposites attract and likes repel...

...until they don't.

Well, you see, protons and electrons ARE attracted to each other~to a point.

Far enough away for a path forty, two,...or even only ONE wavelength long, things are fine.

Get any closer, though, and... disaster!

elemental identity is determined by the number of protons. Also remember that you can get protons to stick to each other—and to neutrons, incidentally, which don't have any charge and really don't care who their neighbors are—if you get them close enough so that the strong nuclear force kicks in.

It is a bit of a challenge to get them that close, though. You've got to get the protons to within about a proton-width apart or they push each other away. That's about a billionth of the width of a human hair. Forcing these guys into such close proximity is not exactly something you can do with your fingers, or even with a really good set of clamps. You need some extreme conditions. First you need to get the protons pretty crowded to begin with, and then you need something that makes them zip around at such colossal speeds that they don't even get a chance to say, "Hey, you've got a positive charge! Get away!" before they run smack into each other. Nothing you'll ever encounter (ideally) will be able to accomplish this task.

Fortunately, there *are* places in the Universe that are able to do this with relative ease. They're called stars. They can heat and compress things beyond your wildest imagination.

So how hot does it need to be? The short answer is "unbelievably hot." The outside of a star like the Sun has a pretty uncomfortable temperature of 11,000 degrees Fahrenheit. But this temperature is freezing cold compared to the temperatures you'd need to convince protons to stick together. Compounding the problem is the complete lack of crowding—an entire roomful of the stuff from the outer part of the Sun would weigh just an ounce.

Burrow further into the Sun and things get a bit toastier and more cramped. It gets so hot, in fact, that the electrons no longer feel a need to stick around their nuclei. ("See, honey? When things get too rough, those opposite charges go their own ways." Eyes roll. "*Yes, Mother.*") Instead of encountering intact atoms, you'd find yourself in a sea of frantic, single electrons and

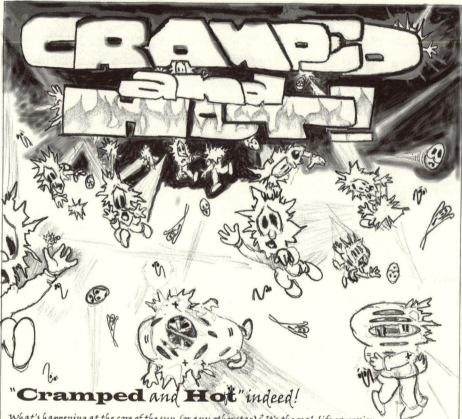

"Cramped and Hot" indeed!

What's happening at the core of the sun (or any other star)? It's the real-life expression of Einstein's equation $E=mc^2$. Once four hydrogen atoms have been squashed and pounded into a helium atom about $\frac{1}{2}$ of one percent of their original mass is **missing**, converted into energy. So, how much energy could that tiny bit of mass amount to? Find your friendly neighborhood **million-gallon** water tower. If you could convert all the water energy **boil all the**

mass in just a thimble-full of (that's about **three grams**) into that would be enough energy to **water in the water tower away.***

*Well, and, starting from 25°C, **27 *other* million-gallon *water towers*** just for good measure. We think demonstrating this would be socially awkward, though.

abandoned atomic nuclei. Temperatures soar to about 15 million degrees Fahrenheit, and material is compressed to the density of lead. But even this part of the star isn't hot enough or crowded enough to get those protons to stick together.

To get to proton-sticking conditions, you need to go to the very heart of the Sun, where temperatures hit a whopping 27 million degrees Fahrenheit and matter is so close together that if you filled a soda can with it, it would weigh over 100 pounds.

As hard as it is to believe, this is even hotter and more crowded than a beach on Labor Day weekend.

Inside the Sun, protons are racing around at over 600,000 miles per hour and bumping into each other constantly, so much so that a single proton will get hit about 10 million times a second. Even so, only about one in every 100 trillion trillion encounters causes two protons to actually stick together.

Tragically, this situation isn't workable for reasons that only subatomic particles can fully appreciate, so one of the protons decides to take on a whole new identity. It loses its positive charge and becomes a neutron. This is probably the nuclear equivalent of a new husband having to get rid of his antler furniture.

So let's take a look at the scoreboard. We started with a couple of separate protons (technically known as hydrogen, even though their electrons are scattered far and wide in the chaos of the star's interior), and now we have something with one proton and one neutron.

'Pretty cool,' you think. But then the realization hits you. 'Wait a minute . . . the nucleus still has just one proton! What a rip-off! You haven't told me how to make a new element! You've just made a heavier piece of hydrogen! I'm returning this book!'

Patience, grasshopper. That's only the first step. Once you have this little particle pair, it's not all that difficult to catch another proton into the mix. After all, they're running into each other constantly. Having the added neutron really helps in this respect, because it has the same type of glue (that strong nuclear force) but isn't annoyed by the presence of another proton. Once you capture another proton, you have a real, live, new element, one with two protons instead of just one. That element is called helium. The process isn't finished, though. A couple of these 2-proton, 1-neutron blobs will invariably hit each other, and there'll be some rearranging that leaves a 2-proton, 2-neutron blob and a couple of freed protons.

Perhaps they weren't willing to give up their antler furniture just yet.

This 2-proton, 2-neutron blob is actually one of the most stable things in the entire Universe, and it serves as the building block for much of the more familiar stuff in your life. For instance, if the conditions are right, you can stick three of these blobs together and make what we know as carbon. Stick four of them together and you get oxygen. Five, and you get neon (so now you can make cool signs in restaurants). Six, and you get magnesium. If you dig out that ancient chemistry book and find the periodic table of the elements, you'll see that lots of the smaller, even-numbered elements are pretty common. Element number 20—calcium—is nothing more than the product of 10 helium nuclei all getting together.

Getting to this level, though, is a bit of a challenge. Stars like the Sun don't have conditions in their hearts extreme enough to do much more than turn unsuspecting hydrogen atoms into helium. Ironically, though, once there's no more hydrogen left to turn into helium, the middle of the star compresses and heats until it *can* succeed in sticking three and even four helium blobs together. And thus carbon and oxygen are born.

Whole fleets of atomic physicists and astrophysicists have worked for decades to figure out the recipes for the elements in the Universe. Some elements, like our favorite, molybdenum, are built gradually in the stellar equivalents of slow cookers that eventually shed their manufactured goods into space. Other elements, like iodine, gold, and platinum, are made in stellar blast furnaces called supernovae, which are the explosive deaths of incredibly massive stars.

The weirder sounding and more difficult the element is to find at your local hardware store, the harder it is for these element factories to make. Most likely you've gone your entire life without knowingly encountering something called dysprosium. It's element number 66, and it makes up all of two out of every 100 billion atoms in the Universe.

Dysprosium can be really useful, though, if you happen to be competing in a spelling bee. Furthermore, it's another thing that sounds like a good excuse to call in sick. "Yeah, that's right." [cough] "Dysprosium." [sniff] "It's pretty rare. I think I'll probably be out at least the rest of the week."

On the other hand, without the common element oxygen (number 8 on the periodic table), which makes up about one out of every 1,000 atoms in the Universe and an astonishing 24% of your atoms, you wouldn't exist.

So that's the basic story for a process known as nuclear fusion. Start with some hydrogen atoms and let stars squeeze them together to make the rest of the elements. You might have noticed, however, that I've completely skirted the issue of where those hydrogen atoms came from in the first place.

Be warned. It's as bizarre as electrons canceling themselves out.

3.3 In the Beginning . . .

Once upon a time, there was this balloon, see, and it was really tiny and filled with intensely short wavelengths of light.

"Oh, great. It's the balloon from the first two chapters again . . . "

And now the balloon is really huge and filled with stretched light with a much, much longer wavelength.

Oh, and there's a whole lot of stuff in it, too. But I've been ignoring that little detail till now. Somehow, though, the Universe went from a place with absolutely *no* stuff to a place brimming with *so much* stuff that we need extra storage units. So, how'd that happen, exactly?

It helps to think of things in reverse. Right now we live in a large, cold Universe. Symbolically, you can think of it as a really big ice cube. Maybe with a penguin nearby, enjoying the weather. So let's go backward in time, shrinking the Universe and making it hotter. What happens to the ice?

"Duh . . . it melts."

Exactly. Now you've got water, and a swimming penguin. But that's no different from ice. It's still the same basic stuff; it's just in a different phase.

Now go back farther in time to a smaller, even hotter Universe. Your puddle of water can't withstand that sort of temperature, so now you've got steam, and a really, really ticked-off penguin. But steam is nothing more than water that's been vaporized. Cool it down and you can make ice again if you want. Once again, the only thing that's changed is its phase.

Increase the temperature even more (and humanely remove the penguin, please). Now the poor little water molecules don't actually have enough strength to hold themselves together. Instead of H_2O, you'll have a bunch of free H's and free O's. It's actually remarkably easy to separate the hydrogen from the oxygen in water. Believe it or not, there are lots of videos on places like YouTube that demonstrate this process, which is called electrolysis. You can even capture the hydrogen in one container and the oxygen in another and then, if you're like my chemistry teacher, see what happens when you put the hydrogen near an open flame. (Think Hindenburg.)

Anyway, for the sake of illustration, let's ignore the O's at this point and look just at the H's. These are just hydrogen atoms, with one proton sitting in the nucleus and a lone electron held at

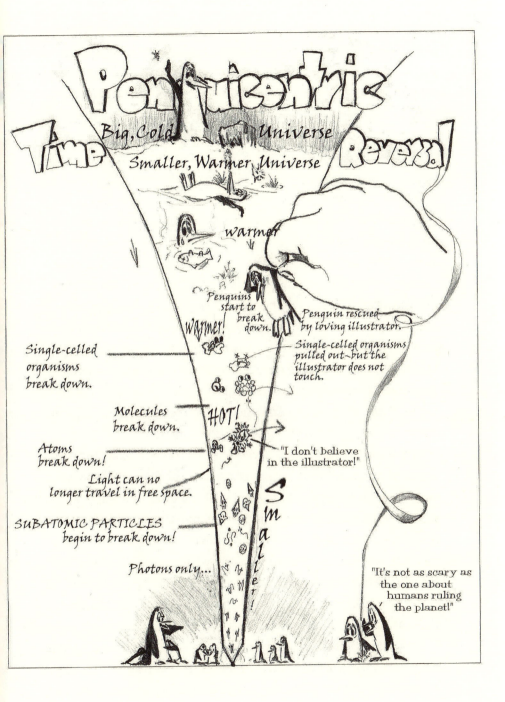

a safe, non-self-canceling distance. It turns out that with a high-enough temperature—like that found in the interior of a star—you'll actually rip the electron from its nucleus. So now you've just got a bunch of free-roaming protons and electrons. But the stuff is still essentially the same. Cool it down and they'll recombine to form hydrogen. Cool it down more (and stop ignoring the oxygen atoms) and you'll make water molecules and ultimately ice.

Now things start getting a little weird. Electrons in and of themselves can handle some pretty nasty conditions, but protons—and neutrons, for that matter—can withstand *only* about 10 trillion degrees before they are ripped to shreds by the intense energy. When they are, they each become a trio of even tinier particles that have the whimsical name of *quarks*. Protons are made from two "up" quarks and a "down" quark, while neutrons are made from two "down" quarks and an "up" quark. These combinations give them their charge (or lack thereof).

And, no, I am not making this up.

There are even charm and strange and top and bottom quarks, but they don't combine to make protons and neutrons. These freaky particles have been revealed in particle accelerators (a.k.a. atom smashers), where physicists race tiny things (typically subatomic particles) toward each other at incredible speeds and see what debris comes flying out when they collide. It's been compared to smashing two watches together and then trying to sort out how they were constructed by looking at the wreckage, but it's the best way physicists have of mimicking the conditions in the Early Universe when things were slamming into each other at unimaginable speeds.

Now things get *really* weird. If you make things even hotter, no actual stuff can withstand the intense energy. "Stuff," it seems, is essentially a frozen form of energy. Einstein's famous equation $E = mc^2$ says so, but very few people actually appreciate what it means.

It's Equation Appreciation Time! All equations equate
two things, or they wouldn't be called equations. In this case
E—Energy—is equivalent to something else. The m stands for
mass, which is a measure of the amount of stuff something pos-
sesses. The c stands for the speed of light, and it's a ginormous
number. If you square a ginormous number, you get an even more
ginormous number. So the essence of this equation is that energy
is equivalent to mass, but there's a ginormous scaling factor.

To think of a scaling factor, think of pennies and dollars.
They're both money, so they measure the same thing, but you
need to scale the dollars by a factor of 100 to figure out the number of pennies. So your equation would read (number of cents)
= 100 x (number of dollars). In $E = mc^2$, you have to scale the
mass by an absolutely colossal number to figure out the energy
equivalent.

In other words, a very small mass is the equivalent of a very
large amount of energy. *Equivalent!* They can be converted back
and forth just like dollars and cents, except that a few specific
physical conditions have to be met first.

So just how much energy is equivalent to how much mass? If
you could somehow figure out a way to convert every scrap of
mass in this book to energy (and, more importantly, contain that
energy), you could provide power to over a million average U.S.
households for a year. You may not believe that energy and mass
are actually two sides of the same coin, but the sad truth is that
humans have engineered a way to capitalize on this fact. About a
pound and a third of mass was converted directly into energy in
the atomic bomb blast of Hiroshima.

That was only a little over one pound of mass, or, to be more
scientifically accurate, a little over a pound of weight as measured on Earth. That's about 600 grams for the more metrically
inclined. Earth itself contains about 6,000,000,000,000,000,000,
000,000,000 grams of stuff. The entire observable Universe has
more like 1 followed by 60 zeros grams of stuff.

None of this stuff was able to survive as mass in that hot Early Universe. Instead it began its existence in the form of energy. That's 10 million billion trillion quadrillion quadrillion (PHEW!) Hiroshima blasts worth of energy contained in a Universe smaller than the head of a pin. And that's only counting the energy that would later become stuff, which, as you'll see later, is a tiny fraction of the original energy. Then there's still all that leftover energy that we now see as microwave radiation permeating all of space.

It's enough to make your brain hemorrhage if you think about it too long.

So don't do that. You have several wonders left to explore, and, anyway, we're trying to pave the way for the creation of hydrogen.

Once that intensely hot, energetic, and tiny Universe began to inflate, things cooled off considerably. The same $E = mc^2$ that we use to create destructive energy from mass worked the opposite way to create mass from the unbelievable amount of available energy. After a fraction of a fraction of a second, things had cooled off enough that electrons could exist. It was almost like reaching a dewpoint. After about a second, protons and neutrons could happily keep all their quarks together. Thus the naked cores of hydrogen atoms were born.

Given the fact that the Universe at this time was hotter and denser than the center of a star, some of the protons and neutrons stuck together to make what would become helium. But things kept getting bigger and cooler, so that process didn't use up all the naked hydrogen nuclei. At final count, 90% of the remaining nuclei were hydrogen, while almost all the rest were helium, with little bitty bits of lithium (element number 3) thrown in for good measure.

The ratios of these lightest elements have remained essentially the same since those first few minutes of the Universe's life. All the heavier elements that have since been manufactured by stars make up less than a tenth of a percent of the atoms in the Universe, but more than a third of your atoms. Still, you and the rest of the Universe share one important trait: most of your atoms have been riding throughout space for nearly 14 billion years.

How's that for making you feel old?

3.4 Making Light of Evil Twins

Of course, there's a complication. Another itsy, bitsy detail that I might have glossed over. The problem I've avoided with the whole conversion of energy to mass thing is that whenever energy (light) turns into particles, it always seems to make both a particle and its evil twin.

Evil twin particles do everything evil twins are supposed to do. They are identical to their twins in virtually every way. In the case of the electron, its evil twin is the positron, something else that sounds like it came straight from a bad sci-fi movie. Positrons are just like electrons except they have a positive charge. They belong to a curious population of things called antimatter. Again, this is science-fiction sounding, but it is very much real stuff. It can be made and even used in physics labs, and some of the more violent places in the Universe appear to be shooting out antimatter.

Just like a proper evil twin, antimatter seems to have an evil purpose in life, namely to seek and destroy "normal" matter. Whenever a positron and an electron get together, for example, they annihilate each other (only without the epic onscreen struggle) and become pure energy.

This is actually the basic mechanism behind sunshine. Remember those protons that had to get rid of their antler furniture—I mean, positive charges—and become neutrons? Those positive charges that were released were actually more than just free-roaming charges, which sound like something on your cell phone

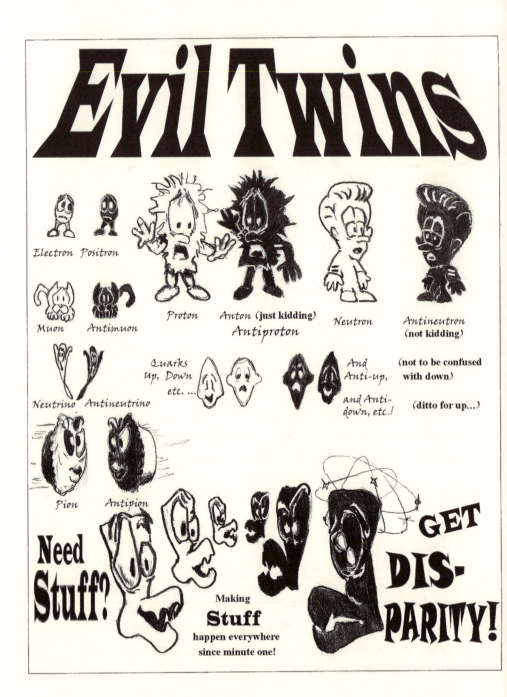

bill. They were the evil twins of electrons, and they were let loose in a dense sea of regular, everyday electrons. Pretty much instantaneously, the evil twins sought out the good twins and they annihilated each other, leaving only light in their place (gamma rays, specifically).

In other words, every time hydrogen fuses into helium in the middle of a star, a certain amount of stuff gets turned into light. Four electrons' worth of stuff, to be more precise. The Sun loses weight and creates light with every single atom of helium it forges in its heart. What's insane is that these reactions are going on at a rate that destroys over 4 million tons of mass every single second! That's like every single car in a major city disappearing *every second*.

That'd be kind of nice during rush hour, come to think of it . . .

Before you run out to say good-bye to our shrinking Sun, consider this: its weight-loss program is equivalent to your losing a single skin cell every thirty years or so. Yes, technically you would be losing weight, but your clothes wouldn't fit any better.

If the Sun could somehow keep going at this rate till all its mass was turned into light (which it can't, but just indulge me for a second), it would last over 10 trillion years. What it *can* do is keep converting hydrogen to helium in its hottest, densest region, while losing a fraction of its mass with every conversion. Losing that stuff makes it . . . well . . . *lighter*. It'll keep pumping out light from this process for about another 5 billion years, and it's all thanks to those evil twins.

3.5 The Good Guys Always Win . . . But Why?

One major puzzle that still remains, though, is why all that light in the hot Early Universe made more good twins than evil twins. As far as physics seems to go, whenever energy becomes mass, the process is forced by a great Universal Accountant to make both the good twin and its evil counterpart. You'd think that these would snuff each other out immediately, leaving lots of light, but still no stuff.

And yet there's clearly lots of stuff and precious little antistuff. There's enough stuff to make you, me, Earth, the solar system, and easily billions and billions of galaxies, each one brimming with billions of suns' worth of material. Clearly something tipped the balance.

But what? And why?

These are deep questions that require physicists to slam particles even harder into each other to reveal answers about how things behaved in the earliest stages of the Universe. You can almost hear them exclaim, "Woohoo! More power!" An experiment in 2008 at Fermi National Accelerator Lab involved something known as a B_s meson (no, ... B_s doesn't stand for that—it stands for bottom and strange, and the official pronunciation is B-sub-s), which consists of, you guessed it, a bottom quark and a strange antiquark.

What ... you didn't guess it?

Anyway, apparently the behavior of the B_s evil twin isn't always an exact mirror of the twin's behavior, and that's a good thing. Physicists have suspected since the 1960s, but never experimentally verified, that perhaps evil twins break apart before they get a chance to kill their good twin targets. Just like with heroes in action movies, the bullet barely misses. The Fermi Lab B_s experiment seems to show this is the case.

What's incredible is that matter apparently needed just the tiniest advantage over antimatter in order to take over the Universe. Only one particle in a billion had to survive being killed by its antiparticle in order for the Universe to make all the stuff there is. That means that 99.9999999% of all the stuff created by the conversion of light to mass was converted right back to light before the Universe was even a minute old!

And given the amount of stuff there is in your closets, you

should probably be pretty thankful that early conditions didn't favor the good twins even more.

SMALL WONDER

What Are You Really Made Of?

Forget frogs and snails and puppy dog tails, or sugar and spice and everything nice. Here's the real recipe for boys and girls. Most of what makes up boys and girls alike is hydrogen (62% of the atoms in the body), oxygen (24%), carbon (12%), and nitrogen (1%). That means Everything Else combines to make up a minuscule 1 out of 100 atoms in your body.

Vital for your teeth and bones, calcium (element number 20) is the next most abundant element, making up about 1 in every 500 of your atoms. This is one of the elements forged by slamming successive helium nuclei together. Coming in a close sixth place is phosphorus (1 in 700 atoms), element number 15, which is found rather abundantly in your teeth, bones, and your very genetic code.

Meanwhile, mergers and explosions of extremely dense stars called white dwarfs are responsible for creating much of the iron (element number 26) that keeps you from being anemic. Even though it's amazingly important, iron makes up only about 7 out of every million of your atoms. Another important, but not very abundant, element is iodine (element number 53), which is manufactured and blasted into space by massive exploding stars. It's absolutely vital for proper thyroid function, but you'd have to sift through more than 100 million of your atoms before you'd run into a single iodine atom.

Some day when you have some time (perhaps one of those 1,100-hour days that we'll get in 5 billion years) check out the list of things in your multivitamin and ponder where those ingredients came from. The 75 millionths of a gram of molybdenum that you take each day to make enzymes for maintaining a healthy metabolism came from stars not all that different from the Sun. They spent billions of years adding protons and neutrons together, ultimately sending their creations into the rest of the Universe. Eventually these and other stellar-manufactured goods came together in what would become our Sun and solar system and, ultimately, you.

And here you are, 7 trillion quadrillion atoms painstakingly made by the Universe. You are, in fact, the Universe thinking about itself.

CHAPTER 4

Gravity

*Gravity is a contributing factor in nearly 73 percent
of all accidents involving falling objects.*
—Dave Barry

"Uh-oh!"

Anyone in the vicinity of a small child immediately thinks of one thing when hearing those syllables: gravity. Thanks to gravity, something that was on the table is now on the floor, and most likely in need of cleaning up.

Whenever it catches our attention, gravity seems to be less a wonder of the Universe than a huge inconvenience. The glass fell off the counter and broke because of gravity. Those numbers on the scale keep inching up because of gravity. Can't remotely make a slam-dunk into a regulation basketball goal? Blame gravity again. Even the cat sleeping on your feet becomes completely immovable in the middle of the night because it taps into some kind of extra-dimensional feline super-gravity (okay, . . . not really).

At some point in your life, after an "Uh-oh" that signals the need for another cleanup in the kitchen, a frustrated kid will ask you why things have to fall. Equally frustrated, you'll sigh and respond, "Gravity." You don't care about natural forces right now. You just want to know the best way to get grape juice stains out of grout.

But one-word answers are never satisfactory to a child, so the next thing you'll hear is, "What's grabbity?"

A much better word for the phenomenon, that's what. "Grabbity" is something that just reaches out to "grab it"—as long as "it" is breakable—and flings it to the floor. But what *is* gravity? More important, what interpretive dances are you going to have to perform *this* time?

4.1 A Penny for Your Thoughts?

Just about everyone has some good intuition about gravity. Unfortunately, just about everyone also has some wretched misconceptions about gravity. The worst part is that almost nobody knows where the good intuition ends and wretched misconceptions begin.

So let's start small. With your arms stretched out to your sides, hold an object in each hand. In your right hand, a penny. In your left hand, a copy of *Gone with the Wind* or whatever it was you finished reading in the doctor's office when figuring out the puzzle of planetary spin rates.

Now hold them there for 30 seconds and ask yourself which one has a larger force of gravity on it.

There are a few possible answers here, and the more science classes you've had, the weirder your answers will seem to a small child. The first possible answer is, "Duh!" as you let your left hand drop to your side, sending *Gone with the Wind* to the floor. Frankly, you don't give a damn. You just want the pain to stop. The second possible answer is that since neither of the objects is actually falling, there must not be *any* force of gravity on them. Another possible answer is that the force of gravity is the same on both of the objects. After a scrunchy-nosed look of perplexity from the child, you simply demonstrate your answer by dropping the book and the penny from the same height.

The child will probably be amazed that they both hit the floor

Mom! Can't we watch the Early Universe instead?

at the same time, but not quite as amazed as astronaut Commander David Scott of the *Apollo 15* mission seemed to be when he dropped a feather and a hammer on the Moon and they hit at the same time. (See http://nssdc.gsfc.nasa.gov/planetary/lunar/apollo_15_feather_drop.html to watch this in action; see also www.sevenwondersoftheuniverse.com for links to more cool astronaut videos.)

But does that really mean that Earth is tugging equally hard on the book and the penny? Your arms tell you one thing ("Ouch . . . book . . . heavy . . ."), and your eyes (and probably some science classes) tell you another. Oddly, Commander Scott's Earthly intuition about what hammers and feathers ought to do when they're dropped side by side on the Moon told him another thing entirely, even though his rational side knew the physics he'd learned *had* to be right or he wouldn't have been standing there to do the demonstration!

The problem with gravity is that there's just so much else tangled up with it in our brains. We think of falling things, but almost none of us has been to a place where falling isn't affected by something else, like air resistance. (If you haven't already, you really need to take a break and see the video of the hammer and feather falling together. No, seriously. Right now.) We think of the specific motions but don't stop to think of how motions and forces are related. And since we're stuck here on Earth, we almost never think about gravity elsewhere in the Universe.

In short, we have serious issues with gravity, and not just around the holidays.

So let's stick to Earth for now. (Not that we have the choice, because, after all, gravity is pretty much forcing us to stick to Earth.) In fact, let's stick to the book and penny example. To disentangle motions from forces, try the following. Set both the book and the penny on a table. Then flick the penny as hard as you can and watch it sail off the table and across the room.

Now flick the book as hard as you can.

"Are you *nuts*?!" you protest. "That'll hurt my fingernail!"

Exactly. Worse yet, the book won't go anywhere. That's lesson one on forces (a hard flick on each object) and resulting motions (sailing across the room versus laughing at your feeble efforts). The thing about the penny is that it just doesn't take much "oomph" to get it to do something dramatic. The book, on the other hand, is going to take a much harder shove if you want to send it across the room.

Now turn this demonstration 90 degrees and think about how much "oomph" Earth has to muster to pull these things to the ground. When you drop them, the book and penny have the same motion—they speed up side by side till they hit the ground. But, as you know with the demonstration on the table, if you want a book to do the same thing as a penny, you've got to provide a whole lot more "oomph." Somehow Earth compensates for the extra mass of *Gone with the Wind* by pulling on it that much harder. But why? And how? And what's so special about Earth, anyway, that makes it pull things toward it?

4.2 Earth, the 6 Trillion Trillion Kilogram Weakling

Over two millennia ago, the ancient Greek philosopher-scientist-politician-teacher-poet-ethicist Aristotle (whose business card was probably pretty sizable) puzzled over the attraction that things seem to have for Earth and suggested that everything just has a natural tendency to go where it's supposed to go.

Clearly the man never stepped foot into my house, where the "natural" place for most stuff is in chaotic piles atop desks and tables.

In Aristotle's system of nature, earthy things like rocks (and apparently breakable stuff) tended to head toward Earth, which was sort of the central anchor of the entire Universe. Watery things flowed to the water, the level above Earth. Airy things went into the air, which was above the water. Fiery things like smoke headed upwards above even the air. And those heavenly things like the Sun, Moon, and stars were in their proper ethereal place above everything else.

Then there were combos. In this philosophy, a feather was somewhat earthy, somewhat airy, which is why it drifts and floats on its way to the ground. The poor thing was suffering an identity crisis the whole time, wondering whether to ride the wind or settle to Earth. Apparently small children in ancient Greek times could get away with telling their parents that it was just time for the pottery to return to its natural position in the Universe.

In later, more sophisticated days, kids weren't able to get away with such nonsense. But they could eventually appeal to this mysterious "gravity," a weighty word that came into use around 500 years ago. With the common introduction of the word, scientists were compelled to hijack it for their own purposes, so now any good dictionary will give you at least two definitions of gravity, one of which is scientific. As you read this, some physics teacher somewhere is making a joke about the "gravity of the situation," hoping against hope that the students will appreciate the pun.

They won't. They'll be too busy trying to memorize the equation for Isaac Newton's 1687 Law of Universal Gravitation, because in the 2,000 years separating Aristotle from Newton, plenty came to be understood and mathematized about this strange interaction between the stuff in the Universe.

If you speak mathematics, you'll be able to translate the official algebraic equation defining the force of gravity:

$$F_g = G \, m_1 m_2 / r^2$$

If you don't speak mathematics, I might as well have let my cat step across the keyboard for a while. So let me try to put that into English.

One thing this equation tells us is that the force of gravity is stunningly weak.

Yes, I know. That's a tough sell the weekend after Thanksgiving, but it really is. Think about that balloon again. No, not the one with the ant traipsing aimlessly along (we'll need that one later), but the one that made the child's hair stand up on end. That same balloon, with its newly acquired charge, can be stuck to a wall, where it'll hang quite happily for a while, blithely disobeying Earth's persistent orders to fall to the floor where it "belongs." A rough estimate puts the number of clingy electrons at a few hundred billion, give or take a factor of 10, all making use of the attraction between opposite charges.

We've already seen just how pathetic *that* attraction is compared to the glue holding two protons together, and yet Earth—formidable Earth, with 6 trillion trillion kilograms of stuff, all trying valiantly to make the balloon return to its "natural state" on the ground—is powerless to overcome that attraction until the balloon's charge dissipates a bit.

In the grand scheme of Universal attraction, gravity comes in dead last to the strong nuclear force, the electrostatic forces between charges, magnetism, and something called the weak nuclear force, which is only weak when compared to those other forces, but is 10 trillion trillion times stronger than gravity.

What gravity *does* have going for it, though, is its reach. There is no fine print that says, "This equation good only for things on Earth." On the other hand, the insanely strong "strong force" engages only within the breadth of an atomic nucleus; the weak force has an even punier range.

Gravity, although weak, knows no bounds. But it does diminish in strength as the separation between things gets larger.

Other than telling us of gravity's feeble grip and incredible range, the equation above also tells math-speakers a few other things that come as a surprise to most people. One is that it's not just Earth that tugs on you. Any time you have two or more

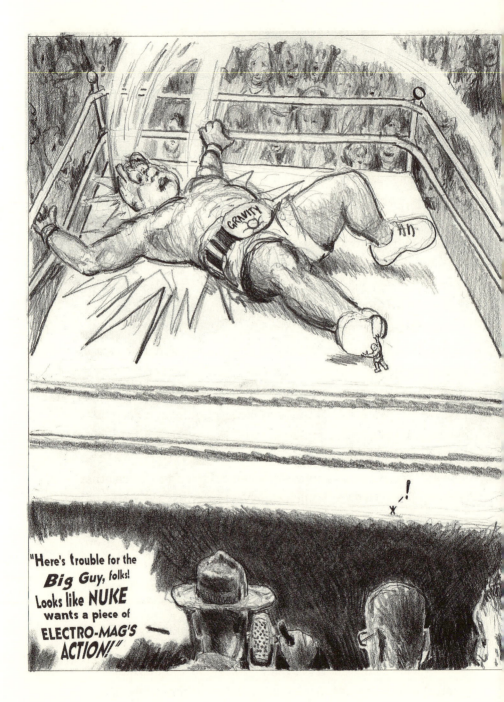

pieces of stuff (or, more precisely, two masses, which is what the
m's represent), there's a gravitational attraction between them.
Technically, there's a physical attraction between you and this
book. That might sound kind of disturbing, but it's true. There's
also an attraction between a pen on your desk and a sofa in a
hotel in Rome, and between the copier at your office and the
technician cursing at it. In fact, all the stuff in the Universe is
gravitationally attracted to all the other stuff in the Universe, but
because gravity is so phenomenally weak, you don't really notice
the effect unless one of those pieces of stuff is really enormous,
like Earth.

Or unless you're an incredibly patient physicist who can set
up a highly sensitive experiment in your lab. Designed by John
Michell, and later bequeathed to Francis Wollaston, there is such
an experiment that is now universally known as the Cavendish
experiment. Naturally. But this one actually makes sense because
Henry Cavendish is the one who finally managed to carry out the
experiment in 1797 after the others had either died (in Michell's
case) or decided they had More Important Things to do (in the
case of Wollaston).

What Michell originally had in mind was pretty weighty: the
entire world, in fact. And he was going to weigh the world by seeing how hard a couple of fixed 350-pound lead balls pulled on a
couple of 1.6-pound lead balls when they were attached to either
end of a beam that itself was suspended by a thin wire. Once he
had this answer, he would simply compare it to the strength of
Earth's tug on one of the small balls and work out the appropriate ratios using Newton's equation.

Just for Michell to come up with the convoluted idea for this
experiment was pretty impressive (to see just how convoluted
the idea is, do a quick YouTube search on the Cavendish experiment), but believing he could measure this was bordering on
insane. A century before, the genius Isaac Newton himself said
such measurements couldn't be done because the tiniest distur-

bance (a fly burping in the room, for instance) would invalidate the results. But Cavendish was a detail-oriented kind of fellow, so what he did was set up his experiment in a sealed room and make his measurements by looking through a telescope set up in an entirely different room. At one point, a minute temperature difference between regions inside the closed chamber set up subtle air currents that ruined everything. A nearby horse-drawn carriage would do even more damage. For ten months, Cavendish repeated and repeated and repeated his experiment, trying to eliminate all possible disturbances and getting lots of practice at reporting errors.

All this time, Wollaston was probably feeling pretty smart for passing the equipment along.

But Cavendish ultimately succeeded in measuring the minuscule motions in Michell's apparatus and, with a bit of reasoning gymnastics, he was able to declare quite accurately the density and, by extension, the overall mass of Earth. As an added bonus, this measurement ultimately led to a more precise formulation of gravity that could be applied not just to things on Earth, but to everything in the Universe.

Take the Moon, for instance. It's got mass and so do you, and all this new information can help you figure out how hard the Moon would tug on you if you were standing there. The answer: a whole lot less than Earth tugs on you when you're standing on Earth. If you want to lose *lots* of weight—over 80% of it, in fact— all you have to do is travel to the Moon. If you weigh 150 pounds here on Earth, you'll weigh a mere 25 pounds on the Moon. You won't look any different, though, if that's what you're after. Your *stuff* hasn't changed, unless you spent the entire journey to the Moon suffering from severe motion sickness, in which case you probably are made of a bit less stuff. Ignoring that rather disgusting possibility, the most dramatic change is the amount of stuff in the thing you're standing on. The Moon has only about 1% the mass of Earth. Another thing that's changed is your distance—the

r on the bottom of the equation—from the thing you're standing on (see "Small Wonder: How to Lose Weight and Keep It Off,
Guaranteed!").

"Wait a minute . . . I'm standing *on* Earth. Shouldn't that make
the distance zero?"

You'd think so, but if you try to get your calculator to divide
anything by zero, it'll just get annoyed. It's not a matter of calculator convenience, though. Earth isn't just a single mass. It's lots
of little ones. Some are right under your feet. Some are an entire
planet away. If you want to figure out exactly how hard Earth is
pulling on you, you have to add up all of the jillions of gravitational tugs between all of your atoms and all of Earth's atoms.

On the other hand, you could just take advantage of another
thing Isaac Newton invented in his spare time and use the mathematical tool called calculus, which consists largely of lots of
shortcuts for adding up jillions of things without all the actual
adding.

Or, an even more preferable approach for most people is just to
take my word for it. Once you get done adding up the individual
attraction between all your atoms and all Earth's atoms, it is as
though the entire Earth is squished into a pinpoint at its very
center.

In other words, you—even though you might be standing at
sea level—are 4,000 miles from Earth, at least for the sake of
the math. If you're standing on the Moon, you're a hair over
1,000 miles from its center. Between its lower mass and smaller
size, the Moon conspires to make you weigh about a sixth your
Earthly weight, which can make things really fun.

Even with incredibly massive spacesuits on, Apollo astronauts
were able to perform amazing jumping feats and occasionally
not-so-graceful falls as their bodies attempted to overcompensate
for the forces they felt. Inexplicably, despite the easily download-
able videos of the various astronauts bounding along the lunar
surface, throwing things, playing golf, falling over backwards,

and dropping hammers and feathers (just check the Internet), a huge percentage of people will assert that there is *no* gravity on the Moon. "Then why do the astronauts come back down when they jump up?" someone once asked a denier of lunar gravity. "Because they have really heavy boots," declared the denier with confidence, which leads one to wonder . . . what makes the boots heavy if it's not gravity?

Technically, though, these people are right about one thing. Gravity isn't something that just oozes out of objects. It's an interaction *between* things, like a tug-of-war. The Moon, sitting there alone, minding its own business, doesn't wield a force of gravity. Neither does Earth, for that matter, but you'd be hard-pressed to find someone who'd say there's no gravity on Earth. It's only when some other thing—a second m—factors into the equation, so to speak, that the two will tug on each other with this strange force that can act over vast distances.

The really bizarre thing is that, given the nature of multiplication, if you compute the force of gravity that Earth exerts on you (or, if you don't feel like manipulating an algebraic equation, just step on the scale), it turns out to be the exact same number as the force of gravity that *you* exert on Earth! It just seems like you're the only one being pulled on, which is why we tend to think that Earth has all this gravity and we're just innocent victims of it.

The problem is your mass. You don't have much of it, whereas Earth out-masses the average adult by a factor of 100 billion trillion. Ten trillion freight trains out-mass a fly by about that amount, too. So what has the most drastic change in its motion when just a single freight train runs into a fly?

Exactly.

Technically—*technically*—a fly splatting on a freight train's windshield will do something to the motion of the train, but it won't be noticeable. *Technically*, when you hop off a chair, Earth is pulled upwards by your mutual gravitational attraction and meets you somewhere in the middle. But it's not measurable. A favorite problem given out by physics teachers is to compute how far Earth would move if everyone in China hopped off a chair at the same time. It turns out to be less than a proton-width. The problem is that it takes a really huge tug to get Earth to do anything.

It's kind of like a teenager in that respect.

Nevertheless, truly snarky kids these days could accurately get away with saying that the glass didn't fall onto the floor. The floor came up to meet the glass.

In any event, with the precise mathematical description of gravity, centuries of introductory physics students have been forced to compute the gravitational attraction between their professor and a fire hydrant down the street, all the while not having the faintest notion of what gravity *is*. As in the case of light, the equation will tell you how gravity *behaves*. It behaves like a force of attraction between all the masses in the entire Universe.

But if that's the case, then why doesn't everything come crashing together?

4.3 Of Apples and Orbits and Confused Astronauts

This very question puzzled Isaac Newton. Why, he wondered, did gravity make an apple fall to Earth from a tree and yet (thankfully) not pull the Moon to the ground? Did gravity just turn off at some point? If so, where? Or was there some other force that kept the Moon at bay?

The answer came in something known as a thought experiment, which is an experiment that he couldn't actually set up and perform, but that he could demonstrate in logical steps. It took only about 300 years to make his idea a reality.

The cool thing about thought experiments is that they're pretty easy to explain to a child without interpretive dances of any kind. To understand how apples fall and the Moon seemingly doesn't, just imagine a baseball and a really, really tall and amazingly powerful baseball pitcher. ("Is it NoOne Rhyme again?" the child asks, because she remembers the autographed baseball from chapter 1, to which your response should be "Um, sure," because you've slept since then.)

Imagine that the pitcher simply lets go of the baseball. What does it do?

The child will wonder why big people are so obtuse as she answers, "It falls."

This doesn't come as any surprise to a child because, after all, "grabbity" got hold of it the same way it gets hold of breakable things.

Now imagine that the pitcher lightly lobs the baseball in the park. Again, the baseball will fall to the ground, but it'll travel outward a bit before it hits. On tiny scales like this, the ground is essentially flat, so the baseball lands at the same level as Mr. Rhyme's feet.

The ground isn't really flat, though, and this is where strange things start happening (and why we need our pitcher to be really, really tall). If the baseball is thrown hard enough, it'll still hit the ground, but as you can tell from the handy illustration highlighting Mr. Rhyme's short career, it actually lands *below* the pitcher's feet. There's no denying that it's falling the whole time, but it manages to take an intercontinental flight before hitting.

Now if you happen to live where the thing landed, your sense of "down" is very different from the "down" of the pitcher. Your "down" is actually pointing to the right in the diagram. In fact, no matter where you are on Earth, your "down" is pointing toward Earth's center because that's the direction Earth is tugging you. This explains why the folks on the South Pole don't just go streaming off the bottom of the planet. Our idea of "up" and "down" is completely artificial when it comes to gravity on

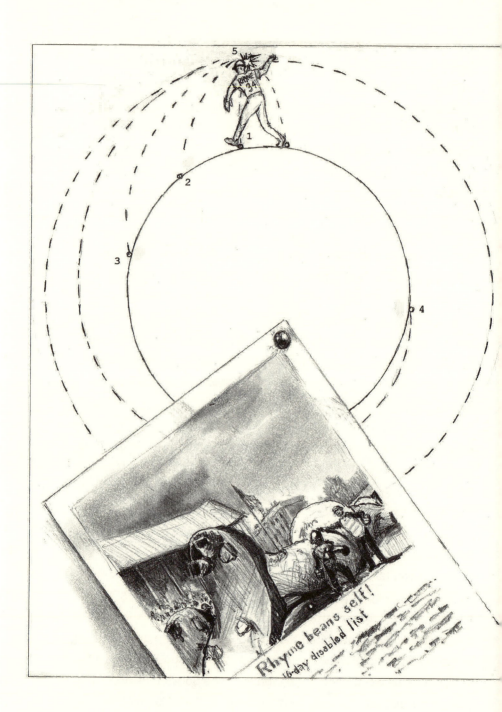

a global scale and exists only because someone, somewhere, decided that the North Pole represented the top. That'd mean that most of us are just clinging to Earth's sides.

Our pitcher isn't satisfied with simply tossing the baseball into another continent, though. He winds up, daring anyone to hit the fastball that's about to come at him, and throws as hard as possible. At first, Earth tries to pull the baseball "down" (that is, toward his feet), but the baseball is moving at a pretty good clip to the left and can't be bothered to hit the ground just yet. Then Earth tries to pull the baseball sort of "down and to the right," but again, the baseball is moving pretty fast, so it doesn't land. Instead of making it hit the ground, gravity alters the baseball's direction slightly. Now it's headed "downward," still moving fast, still not hitting the ground, and Earth is trying to pull it straight right. This sort of compromise between the direction the baseball's moving and the direction gravity is trying to pull it goes on till finally, unbeknownst to our pitcher, the baseball comes full circle and smacks him in the back of the head. With its speed completely sapped by the collision, the baseball will finally fall to the ground.

After being beaned by the world's fastest fastball, an unconscious Mr. Rhyme will probably do the same.

Strangely enough, the baseball has been trying its darnedest for the entire trip to fall "down" to the ground, but the actual direction of "down" keeps changing because of the roundness of Earth. On the next throw, if NoOne Rhyme remembers to duck, the baseball will just keep going around and around and around . . . always falling, but never hitting.

Obviously we don't launch baseballs like this, but the idea still works. If you can get something going fast enough "sideways," it will keep missing the ground as gravity tries to pull it down. "Fast enough" turns out to be about 18,000 miles per hour. This is a bit faster than a pitcher can zing a baseball, but not impossible for a couple of million-pound solid rocket boosters and an additional fuel tank containing 140,000 pounds of liquid hydrogen, nearly 400,000 pounds of liquid oxygen, and some serious thrusters. That kind of equipment is what propelled the Space Shuttle from zero to 18,000 miles per hour in just over 8 minutes. Once you get the speed and direction right, you can shut everything off and just naturally coast around the planet.

Here's the really disturbing part. Things that are launched into orbit are, like the baseball, always falling but never hitting. Being in orbit means to be in free fall, a state that gives the *illusion* of zero gravity but is, in fact, a testament to gravity's hold. If gravity just "turned off" at some point in space, nothing would be making astronauts and their space ships perpetually circle Earth. In fact, astronauts in, say, the International Space Station are being pulled toward Earth with almost the exact tug that they'd get if they were standing on Earth's surface. They simply don't feel their weight, just as a person falling off a roof doesn't feel his own weight.

Not till he hits the ground, anyway.

What this means is that astronauts are constantly experiencing the sensation of falling.

Constantly.

24/7.

Like an endless, nightmarish free-fall ride at an amusement park. For most of us mortals, this sensation thankfully ends after a couple of seconds on a "fun" ride, but some astronauts in orbit get to feel this way for days, weeks, or months. The current record holder for this nauseating endurance test is cosmonaut Valery Polyakov, who spent a whopping 438 days *in a row* in the Russian space station MIR as it fell around Earth.

Of course, after a while the body gets used to this sensation. Once you get past the idea that your hair is supposed to lie against your head and that "down" actually exists (and that days last longer than 90 minutes!), your brain adjusts. In fact, for astronauts who spend extended periods of time in free fall, returning to a place where things actually *do* fall to the floor is quite a shock. It's not unusual for an astronaut back on terra firma to let go of objects in midair fully expecting them to stay put.

This is, no doubt, followed by an "uh-oh" and a cleanup in the kitchen.

4.4 But Wait . . . There's More!

Back in Newton's day, a single equation that simultaneously explained both falling objects and things in orbit had amazing potential. After all, it's not completely obvious that the same thing holding you to the floor is also holding the Moon in orbit around Earth, especially given that the Moon is a good 240,000 miles from here. More impressive is that the Sun manages to keep poor, reclassified Pluto in orbit, and those two are nearly 4 billion miles apart. And the entire Milky Way does a great job of keeping our pal, Pal 4 (you remember, . . . the globular cluster with the epic nighttime view of a rising Galaxy?), in attendance, even though Pal 4 is over 2 quintillion miles from the center of the Galaxy.

With a relatively simple formula, scientists were finally able to understand the driving force behind the orbits of everything in the solar system, and, as it turns out, everything that orbited anything. You could even use it to predict the existence of previously undiscovered objects. All you have to do is add up the gravitational influences of every known object in the solar system and then see if the resulting motions play out properly.

In the case of Uranus-né-George, things were a bit off. The discrepancy between its actual orbital motion and its predicted orbital motion, growing bit by bit throughout the early 1800s, left scientists with two possibilities: (1) everyone's favorite equation was actually not as universal as they had hoped; or (2) some undiscovered world lay in the outer reaches of the solar system. Not wanting to ditch an equation that had worked so beautifully for 150 years, physicists and astronomers worked to figure out where this undiscovered world would have to be.

After a few quick computations (by "quick" I mean over four years of constant devotion to the problem), two independent astronomer/mathematicians, one French and one British, arrived at the answer on almost the same day in 1845. The true test of Newton's gravity hinged on actually finding an object at the mathematically predicted location. A small soap opera ensued where the visual confirmation was thwarted by such trivia as early dinner times, sprained ankles, and lost letters. Finally, in 1846 a German astronomer was told to point his telescope to a particular place in the sky, and he immediately spotted the new planet Neptune. (Turns out that people had probably seen it before but just never made note of it. But that's another story . . .)

It was a stunning victory for such a little equation! Things were definitely looking good for Newton's gravity, and by the late 1800s scientists were genuinely optimistic that everything that could be known scientifically was known. The only things remaining were details—pinning down the exact value for that big G in Newton's equation, for instance. But the "big picture" seemed to be taken care of.

No, really. That's what people were thinking. After all, so much was explained just by gravity. Want to understand those pesky tides that so irked Galileo and countless generations before him? Nothin' but gravity. Specifically, the fact that gravity pulls more strongly when objects are close together and more weakly when they're farther apart. Since you've probably noticed that Earth is not, in fact, compressed into a tiny point 4,000 miles below your feet, part of our planet is always 8,000 miles closer to the Moon than another part of our planet. That difference in distance means the Moon-ward side of our planet always gets a bit more of a gravitational tug from the Moon than the opposite side. The most noticeable effect is found in the ocean levels, since water can actually flow toward the harder tug. Less noticeable is that even the level of the ground that you're standing on can rise and fall by about a foot with the overhead passage of the Moon.

These gravitational tidal effects aren't unique to Earth at all. Anywhere there's gravity and objects with some size to them (in other words, pretty much everywhere), there are tidal effects, and sometimes they destroy more than your sand castles. For instance, in 1994, Comet Shoemaker-Levy 9 passed a bit too close to our planetary king, Jupiter. The extra gravitational tug on its Jupiter side was enough to rip the hapless comet apart into twenty-one baby comets, which then spiraled into Jupiter itself. Jupiter swallowed them all without so much as a burp, but they did leave their mark in Jupiter's upper clouds for a while.

On the less destructive side, gravity also answers almost all the questions about how we got here. Want to form a solar system complete with shining star in the middle? Just start with a ginormous cloud of gas and dust (where that cloud came from was sort of glossed over in the papers of the nineteenth century) and add a heaping helping of time and gravity. Gravity has no choice but to try to pull everything to the middle of the cloud, and just as in the experiment you did on the spinning doctor's stool, the whole cloud has no choice but to spin faster and faster. This spinning up will essentially cause a pancake of stuff to be slung outward like a spinning ice-skater's hair. In just 30 million short years, give or take, you've got the raw ingredients for planets, complete with the flatness of the real solar system and the correct direction for all the orbits. Static electricity starts collecting bits and pieces of stuff together inside that pancake, and gravity does the rest.

Gravity even seemed to explain why the Sun was shining. As the gas and dust from that ginormous cloud collapsed inward, all the atoms and molecules sped up. As they sped toward the middle, the center of the cloud got hotter and denser. In fact, it would ultimately become so hot that it would pump out as much energy as the Sun. The same Lord Kelvin—whose real name, you'll recall, was William Thomson—actually computed how long the resulting bright, hot, central object could remain bright and hot. According to his calculations, the Sun could shine for upwards of 100 million years doing nothing but sapping energy from the gravitational collapse into a hot ball in the center of our solar system.

That is, he cautions in his 1862 paper on the topic, "Unless sources now unknown to us are prepared in the great storehouse of creation."

It's funny how whenever someone says something like that, the Universe always obliges. Just a few decades later, Einstein would hit upon the true source of the Sun's energy, something that could keep it shining for 100 times that long. But you already knew that because you read all about nuclear fusion in chapter 3.

Still, without gravity to get things going, there would be no hot, life-giving orb in the solar system shining under any power—fusion or otherwise—or any planets from which we could admire it.

It's not a stretch to say that we owe our very existence to gravity.

It's just a shame that we don't really understand it.

4.5 Surprise! Gravity Isn't a Force!

By the early 1900s, everyone was delighted with this wonderful force of gravity and all the things it could do. It had never really occurred to anyone to think of gravity as anything but a force, something that affects motions of things with mass, and with over 200 years of testing and victories under its belt, Newton's Law of Universal Gravitation was looking unstoppable. That's why it became one of Newton's Laws, not Newton's Suggestion.

There were some disturbing problems left over, though, particularly in the inner solar system, where the planet Mercury refused to behave. In fact, Urbain Le Verrier, one of the mathematician/astronomers who was set on the task of computing the location of the as-yet-undiscovered Neptune, noticed in 1859 that Mercury's motion was just a skooch different from what it should be if Newton's gravitational equation was correct. But with the stunning victory of Newton's equation in dealing with the slight problem with the orbit of Uranus, Le Verrier was pretty confident that Mercury's aberrant motion would be similarly dispensed with.

Unfortunately, the problem only got worse, and for the next six decades, astronomers and mathematicians trotted out all sorts of bizarre explanations for Mercury's motion. What Mercury was doing was tracing out a sort of Spirograph pattern in its orbit. Now this in and of itself is not the problem, because Newton's Law of Universal Gravitation predicts that such loop-di-looping should happen.

The problem lay in the size of the shift between each loop.

Each of Mercury's orbits is actually shifted a hair of an angle from the previous one by an amount that only a true Type A personality would ever have discovered. Were it really a Spirograph pattern drawn by a kid, it would contain over a million loops, each loop taking three Earth months to draw, with each loop shifted a tiny bit from its predecessors. For Mercury to complete the pattern and return to its starting loop actually takes some 260,000 Earth years, so imagine how little of the pattern could get drawn in just the few decades that observers were paying attention! Yet, this is what astronomers were working with.

Unfortunately, for Newton's law of gravity to be completely correct, the pattern would repeat itself only every 280,000 or so years. Lots of possibilities were suggested for this discrepancy, and not a single one of them hinted that we hadn't been watching the loops long enough to be sure. Perhaps, some thought, there

was another undiscovered planet (it would have been named
Vulcan, by the way, which makes one wonder if Spock would still
have called it home), or maybe a bunch of unseen dust around
the Sun (despite the fact that there was absolutely no observa-
tional evidence for this), or even some additional "tweak" that
was required of Newton's equations.

What turned out to be the best of all explanations was, naturally, the most bizarre. And it would take a mind that didn't just
think outside the box to figure it out, but a mind that didn't even
really know there was a box to begin with.

For our purposes, though, let's think inside the box. So go get
one. A laundry basket or even a child's wagon works well, too.
While you're at it, grab that autographed baseball or some other
good rolling object.

Got everything?

Good. You are now ready to explore Einstein's General Theory
of Relativity.

In interpretive dance!

Place a ball in the box/basket/wagon and pull it along with a
pretty steady forward motion. Now make a left turn. What happens to the ball?

Yes, it seems like yet another "Duh" moment masquerading as
science. Stick with me, though.

If you play with the ball-box setup for a while, you'll no doubt
notice that when you turn left, the ball hits the right side of the
box. When you turn right, the ball hits the left side of the box.
When you slow down, the ball hits the front. When you speed up,
the ball is pinned to the back.

But why?

It's doubtful that you'll spout off Newton's laws of motion at
this point, but the answer is pretty clear. The ball is happily
trucking along in a straight line at a steady speed with the box
and would like to continue doing precisely what it's doing. The

problem is that the thing it's riding in is occasionally changing speed or direction. As long as you keep a steady forward motion, the ball just goes along for the ride. Once you deviate from a steady forward motion, the ball runs into some part of the box.

In a million years you'd probably never hear anyone explain the observation this way: "Actually it happened because there was a gravitational force that suddenly popped into existence and attracted the ball to the side of the box. Furthermore, the gravitational force between a ball in a box and the sides of the box only exists when you change the speed or direction of the box."

That'd be crazy, right?

Einstein thought so, too, but then he took crazy to a whole new level. Our perception of gravity, he said, is nothing more than our being in a box that is perpetually changing its speed. To understand what he's saying, get out of the box for a second and now imagine a windowless rocket ship in space, far, far away from any objects. An astronaut floats around inside because, unlike the free-fall sensation he gets while in orbit, he really is in zero g. Now the thrusters turn on and the rocket accelerates. The astronaut slams to the floor and feels the sensation of "weight" again. But here's the kicker: there's not a force of gravity pulling him to the floor. It's just the fact that the rocket is speeding up, so he sticks to the floor just like the ball will stick to the back of the box when you speed up the box.

Einstein went even further and dared to suggest that there is not a single experiment that this astronaut can perform to let him know whether the "weight" he feels is due to being in a rocket sitting on the launchpad or simply in an accelerating space ship. In one case we'd say there is a force of gravity holding him down. In the other, we'd say it's like the ball-in-the-box demonstration.

And that means that gravity can be thought of in a whole new way. It's not a force at all. It is, in physics parlance, "a noninertial frame of reference."

Whatever it was, the pile of mathematics that followed Einstein's simple analogy explained Mercury's peculiarities to a tee. If this were all it could explain, though, most physicists would probably have dismissed it in favor of dust rings or undiscovered planets or some other thing that actually makes *sense*. The idea that gravity, with over two centuries of experimental verification, was not a force was not one of those things. But Einstein's idea went further than just explaining the motion of Mercury. It predicted something never before observed, specifically, the bending of starlight.

Back to the rocket ship. Imagine there's a tiny window high on the wall of the rocket and that a beam of light comes through. If the rocket is going faster and faster and faster "upwards," then the beam of light will have a very distinct (and, more importantly, calculable) arc downward and hit the opposite wall at a height lower than it came in. If Einstein was right, and if there really is no experiment that can distinguish between this accelerating rocket and a rocket sitting on Earth experiencing Earth's gravity, then gravity should bend light as well.

But that prediction didn't make any sense because light has no mass, and if you plug a zero into one of the m's in Newton's equation for gravity, then there's zero gravitational force to make it bend.

Of course, you know the end of this story. Had Einstein been wrong, then there wouldn't be an entire series of products called Baby Einstein. Why encourage your kid to be a quack scientist with bad hair, after all? But he turned out to be right. Light does bend as it passes near massive objects. The effect was measured in 1919 during a total eclipse of the Sun, and light from stars whose positions were known precisely when the Sun was *not* in the way (i.e., when Earth was on the other side of its orbit) wound up being ever so slightly off. Elegant as this observation was, though, there are those who say the answers were skewed to match Einstein's predictions. Regardless, this so-called gravitational lensing has been observed countless times in the intervening century.

So what does matter do to bend the path of light if it's not some kind of force of gravity?

Our ant's been pretty bored for a while, so imagine, if you will, an ant on a balloon. Actually make it a really big rubber sheet that extends for miles in every direction. The sheet is painted with gridlines, just like a huge piece of graph paper. And bowling balls are placed here and there on the sheet so that they create large dips and stretch the straight gridlines into the dips.

Having been fooled by the balloon already, the ant is bound and determined to keep walking in a perfectly straight line. So he gets on a straight gridline and stays on it. What he doesn't know is that the gridline happens to pass near a bowling ball. Even though the ant stays on the line, he *still* doesn't move in a completely straight line because, unbeknownst to our ant, the line is painted on fabric that is curved and warped by an assortment of masses on its surface.

Our Universe is very much like that. Things with mass, particularly things with lots of mass like stars and galaxies, have the effect of curving the Universal fabric. The fabric of spacetime, it's called. Light behaves like an unsuspecting ant as it tries to follow straight lines, but the lines themselves aren't straight, so light bends around things and gives us a sort of funhouse mirror view of some of the more exotic things out there. In fact, Hubble Space Telescope has taken dozens of pictures of distant galaxies that have had their images distorted and multiplied as their light passes through and around nearer clusters of galaxies. The light traces out the strange warps and dips in spacetime and comes at us sometimes from several directions at once. If you want to see what these wind up looking like, just go to http://hubblesite.org and search for "gravitational lens" and tons of them will pop up for your perusal.

In this concept of gravity, there really isn't some force of gravity holding you to the ground or even pulling the glass to the floor. You're just a little ball resting in the giant dip of a bowling ball. You can try to get out—or send something else out—but it's a bit like trying to roll a ball uphill. Without enough "oomph," the ball will simply roll back down. Now, we don't *witness* some kind of 3-D (or 4-D, as the case may be) dip. What we actually observe is that the ball goes up in the air and then back down to the ground. As it turns out, though, the shape of the dip is exactly what would give rise to Newton's gravity equation.

It's not just falling, though. You can even create stable orbits in these funnel-shaped dips. You might have seen giant funnels called gravity wells set up at a mall or museum. Just put your coins in the launching slot on the top and watch them as they hypnotically orbit around and around and around the central hole. Small children are pretty adept at dropping coins straight in without giving them any kind of "sideways" speed, thus imitating a simple falling action. In either event, the shape of the well imitates a dip in spacetime caused by mass. The biggest difference is that all the friction from the rolling and bumping makes your monetary orbits decay pretty quickly, and before you know it, you've dumped quite a bit of money into a black hole.

Just tell yourself that it's going to a good cause and roll another nickel in there. This is Einstein's General Theory of Relativity you're exploring, after all. It's worth a few more cents.

A huge pressing question remains, though: why should matter affect space in this way? This is one of the most puzzling and exciting questions in physics, and many hopes at getting some answers are pinned upon something called the Large Hadron Collider, a particle accelerator that was scheduled to be up and running by early 2009 but experienced some unforeseen delays and only got on its feet in spring 2010. The "god particle" is what they're hoping to find, and with it the reason behind the existence of both mass and gravity.

And we might finally get the answer to the question, Why do

Flutie doesn't really CATCH mice...

...they just fall into her gravitational well.

things have to fall? After tackling that question, we should be in good shape to answer an even more difficult one, What is the best solvent for getting grape juice stains out of grout?

 ## SMALL WONDER

How to Lose Weight and Keep It Off, Guaranteed!

Now that you know that weight has everything to do with gravity, here's the real secret to losing weight. Just go somewhere else. If you're not satisfied with your Earthly weight, all you need to do is head to the planet Venus, where the combination of the Venusian mass and size would make you weigh only 91% of what you're used to. Okay, so you'll be rather uncomfortably dead, but the scales will read a slightly smaller number. A less nasty place to visit would be Mars, where you'd be a mere 38% of your usual Earthly weight. That means that if you weigh 200 pounds here on Earth, you'd drop to an unbelievably low 76 pounds on Mars! If you really want to drop the pounds, our Moon is more local. A 200-pound person there would weigh in at a mere 33 pounds, and it'd be child's play for a power lifter to pick up a Volkswagen Beetle. If you'd like to be able to pick up that Beetle yourself, you might consider heading to the asteroid Ceres, which is about 500 miles wide. Things weigh just 3% of their Earthly weight there, so the VW would tip the scales at around 45 pounds. After lifting it over your head for a photo op, you could strap on a red cape and leap over tall buildings with a single bound.

Just remember to bring some oxygen.

There are some places you *don't* want to visit if you're interested in losing weight. Just a few are Jupiter, where scales would read 236% of your Earthly weight, Saturn

(106%), and Neptune (113%). Of course it doesn't matter because you can't actually stand on the surface on any of these planets, being gas giants and all . . .

CHAPTER 5

Time

Time flies like an arrow.
Fruit flies like a banana.
—Groucho Marx

You're staring in horror at the field of debris left in your living room—the result of four chapters' worth of demonstrations—when out of the blue comes another impossible question, "What time is it?"

'Oh, great,' you think. 'I'll tell her that it's seven o'clock, and then she'll start asking *why* it's seven o'clock and why we keep time the way we do and what time *is*.' You take a deep breath, steeling yourself for the onslaught of whys. "It's seven o'clock," you say.

"Okay," she says, and skips away.

Your brow wrinkles with confusion. 'Okay? That's all she wanted?' you think. 'She didn't want to know everything there is to know about time?' The confusion morphs into joy. You're off the hook! And the great thing is that she might never ask what time is.

The joy morphs back into confusion. Now *you're* the one who wants to know everything about time, and you have no idea what your clock is even measuring.

Although we kill it and waste it, we never seem to have enough of it. Occasionally it flies, which makes things tough when you're racing against it. A seven-year-old can tell it, but even theoretical physicist John Archibald Wheeler of black hole fame was hard-pressed to define it, stating simply that, "Time is nature's way of keeping everything from happening at once." On the other hand, that exact quotation has been attributed to comedian Woody Allen and to graffiti on a bathroom wall in Austin, Texas. Perhaps time will tell who said it first, but like a child's game of gossip, quotes become more disordered as time passes. Come to think of it, the Universe is like that, too, and it's high time that I try to explain this bizarre wonder called time.

5.1 Got a Second?

Perhaps the most striking thing about time is that we can measure it with amazing precision without really being able to define it physically. Nobel Prize–winning physicist Richard Feynman pointed out that most definitions of time are frustratingly circular. Check out virtually any dictionary and you'll find that time is defined as some period or interval between certain events. Okay, that makes sense. So what's an interval? According to the same dictionary, that's the *time* that passes between certain events.

Great. Even though we didn't manage to define it in our dictionary search, we can be pretty sure that we just wasted it.

Other definitions of time are tied to motion. For example, a second is officially defined as the time required for 9,192,631,770 oscillations of a particular transition of a cesium-133 atom. If I were you, I wouldn't worry about what that actually means. Just be happy knowing that someone's bothered to count them. Anyway, we know that *something* else is going on as things in those cesium atoms are oscillating away, but what exactly that something is, we're not sure. We just declare it some unit of "time."

While we don't usually go around measuring our lives against the oscillations of cesium atoms, this is exactly the type of concept we use all the time. We are drawn to regularities, building our existence around cosmic rhythms that we observe in natural motions.

Think about it. What's a day?

"That's easy," you say. "Twenty-four hours."

Okay . . . so what's an hour?

"Um . . . can I change my answer?"

Furthermore, what sadist decided that there should be 24 of them in a day? It would be so much easier to teach kids how to tell time if we sliced the day into 10 parts instead of 24, wouldn't it? Sure, each hour would be longer, but it would get rid of the whole AM/PM issue. And while we're at it, why not divide the hours into 100 minutes instead of 60? Really now . . . 60? Why such a bizarre number?

Who is responsible for this mess?!

Before we gather up the torches and pitchforks and start hunting down the inventors of our timekeeping system (they're all long dead, by the way), let's get back to that original question. What's a day? And don't succumb to the temptation of declaring it's a certain number of hours or minutes or seconds or any other unit of time. What is *really* going on that tells us that a full day has passed?

It's not as complicated as counting billions of oscillations of a cesium-133 transition, but we still fixate on a rhythmic motion: the Sun appears in our sky; the Sun disappears; and the Sun appears on the horizon again. Pretty much all the inhabitants of Earth—human, animal, plant, and insect—pick up on this rather regular cycle of light and dark. We sleep and wake with the rhythm of Earth's rotation, which at this point in history brings us around to greet the Sun every 794,243,384,900,000 cesium-133 oscillations. Give or take a few billion.

Since our lives are pretty much governed by this cycle, it deserves a name. In the English-speaking world, we call it a day. But a day is so darned long that, unless you're a service technician ("We'll be there to repair your furnace sometime Friday"),

INTERVAL BETWEEN REGULAR EVENTS

PERIOD OF LIGHT'S TRAVEL BETWEEN KNOWN POINTS

OSCILLATIONS CESIUM-133 ATOM

TIME

it has become necessary to break it up into smaller pieces. So we cut the day up into 24 slices.

But, again . . . why 24? The answer to this question depends on whom you ask. On one hand, you'll find those who argue that the answer lies on the hand. While many cultures like to divide things into groups of 10—the typical number of fingers and thumbs humans possess—some scholars have blamed the Babylonians for fixating on the number of *segments* in the fingers. (Note: If you're going to blame someone for something like complicating our timekeeping system, picking a civilization that died out 4,000 to 5,000 years ago is pretty safe.) Using your thumb you can touch 12 different finger segments—3 on each finger. Since you've got two hands, that makes 24 total segments. So perhaps the day should be divided into 24 segments as well: 12 for the daytime and 12 for the night. One hand for daylight hours; one hand for nighttime hours.

Not everyone is satisfied with that explanation, though, so we look farther into the past at the ancient Sumerians of Mesopotamia (about 5,000 to 6,000 years ago), who were apparently big fans of the number 60. Not that they had 60 fingers or toes, of course, but apparently this number was vitally important to them. It could have been the product of the number of finger segments on one hand and the number of digits on the other hand. It could be the number of Sumerians that it took to screw in a light bulb. It could have just had a nice ring to it. Why they loved the number 60 is anyone's guess, and a quick Internet search will convince you that just about anyone who has thought about the question has made a guess. Suffice it to say, 60 was where it was at.

When the Sumerians gave way to the Babylonians, 60 was adopted as *the* all-important number. Babylonian mathematical tablets are filled with complex symbols in a counting base called sexagesimal. To those of us trained in the decimal system, it looks like an attempt by a four-year-old to draw flocks of birds,

but it worked amazingly well for them for centuries and it did ultimately affect much of our geometry.

So if they loved 60 so much, shouldn't there be 60 hours in a day? A truly long-shot explanation starts with a small increment of time: the human heart rate. At rest, a healthy Babylonian man would probably have had a pulse rate of 60 beats per minute. Since the heart was important, and since the number 60 was even *more* important, it's possible that they created a unit of time around 60 heartbeats. It's also possible that this was how quickly Babylonian kids counted out "1 Mesopotamia, 2 Mesopotamia, 3 Mesopotamia . . . 60 Mesopotamia. Ready or not, here I come!" as they played hide-and-seek.

The world may never know.

In either case, it would take about a minute. But why stop there? Sixty of *those* units could be gathered together to make up their own unit of time, which turns out to be an hour. Tragically there are only 24 of those longer units that could fit from sunrise to sunrise, but hey . . . we *did* manage to make use of two 60s along the way.

Like I said, it's a long shot. Anyway, the Babylonians aren't around to confirm or deny this explanation.

Of course, there are plenty of other possible reasons behind the 24-hour day. Ask a mathematician and you'll get the answer that 24 is just a wonderfully divisible number, a perfection no doubt capitalized on by ancient cultures. You can have 1/2 a day (12 hours), 1/3 of a day (8 hours), 1/4 of a day (6 hours), 1/6 of a day (4 hours), 1/8 of a day (3 hours), and 1/12 of a day (2 hours). Clearly having 24 subdivisions in a day makes assigning work shifts and filling out time sheets that much easier.

Ask an astronomer, though, and you'll probably get an astronomical and observable explanation. The day, after all, is based on an observation of something going on in the heavens, something that cultures everywhere have independently hit upon. For that matter, so are the year (the amount of time between succes-

sive "longest" days, or summer solstices) and month (the amount of time for the Moon to cycle through all its phases). Why not the humble hour?

It turns out that there *is* an astronomical phenomenon that is almost exactly an hour in length, and it has to do with the motion of our Moon. Like the Sun, the Moon appears to rise and set because of Earth's rotation. But the Moon is busily going around us as we're spinning on our axis each day and working our way around the Sun each year.

Attempting to visualize how all these motions play out in our Earthly sky is pretty difficult for most people, so order some pizzas and get your friends and family together for another round of "Astronomy in Interpretive Dance." You'll need enough people to play the Sun, Earth, Moon, and several stars. The Sun stands still in the middle of a large room (or, better yet, in the middle of the yard) and holds a lamp. The stars just stand in place. Really, really far away from the Sun—against the walls, fence, or, to be more accurate, somewhere around the city limits.

Earth walks around the Sun *and* spins 365 times for each trip around. Because of the high chance for motion sickness, it might be a bit tough to convince someone to take on the role of Earth, so you might have to do it yourself. That's fine because, after all, *you* are the one who's trying to see what all these things look like from Earth. Meanwhile, the Moon walks around the spinning, moving Earth in such a way that the Sun, Earth, and Moon line up in that order about once every 30 Earth rotations.

Think you can handle all that while making observations of the relative positions of your friends and family members?

A less-daring approach would simply be to look up a good website that animates these motions. There are dozens of them that pop up if you simply type "Sun Moon Earth animation" into any good search engine. Go ahead and do it that way if you like. But you'll just be cheating yourself and your loved ones out of a great opportunity.

Whether you choose the richness of personal experience or opt for the quick Internet animation cop-out, you will probably notice a few things right off the bat. One is that the lamp (or animated Sun) comes into view and goes out of view every time you (Earth) spin. Voilà! A day! Another thing you might notice is that you can actually see that the Moon is doing something quite different from both the Sun *and* the stars. Not just on a monthly time frame, but even from one day to the next. If the Moon is in front of one of your distant stars—Uncle Ted, perhaps—at some point (it's important to call out encouragement to the stars periodically, by the way, or they might get bored and wander off), it'll have moved slightly out of line with that star when you rotate back around on your next "day."

In reality, the Moon appears to move against the distant background stars at a rate that is nearly imperceptible. If you go out on a moonlit evening, it looks like both the Moon and the stars rise and set and traverse the sky in lock step. But if you are an incredibly patient and keen observer, you'll actually notice that the stars seem to trek across the sky at a slightly higher speed than the Moon. Occasionally a star or planet even passes behind the Moon in an event called a lunar occultation, which sounds much more ominous than it is. All it means is that the Moon is eclipsing something other than the Sun. Tons of websites give dates and times and prime viewing locations for lunar occultations throughout the year, the most official of which is run by a group called the International Occultation Timing Association, or IOTA. I suppose it can be said that this organization actually does make an iota of difference.

If you catch one of these occultations just right (they're particularly impressive during a crescent Moon, by the way) you'll see the Moon appear to eat a star or, even more spectacularly, a planet. With any luck, everything is aligned so that the star or planet disappears behind the thickest part of the Moon and has to cross behind the full lunar diameter before popping back out again on the other side.

The amount of time for this entire event is—you guessed it—almost exactly an hour. In an age without Facebook or iPods or electric lights or other distractions to keep them inside at night, ancient peoples would have noticed this sort of thing happening quite often in their dark skies. And given that the Moon held a special place in virtually every culture—entire calendars were based on lunar cycles, rather than solar ones—it probably made sense to note how long it took for the Moon to trek a lunar diameter across the background stars.

And thus the hour as a unit of time was born.

Maybe.

The only problem with this elegant idea is that a complete occultation across the full lunar diameter actually takes about 63 minutes, not 60, and a day can hold only 23 of those blocks of time. But perhaps they liked 24 more than 23 (see mathematician's answer, above). Or perhaps the ancient methods of measuring that time span were a skoochy bit off, an error that's completely understandable. A direct diameter-crossing occultation would be extremely rare and quite difficult to distinguish from one ever so slightly above or below, so maybe they averaged results. Worse still, without a reliable clock, it's hard to accurately time the motion of something, especially when you're trying to make that something the basis of your timekeeping in the first place!

In any event, that's the story many astronomers stick to, and there aren't any Sumerians or Babylonians around to confirm or deny that one, either. But no matter how they settled on the length of the hour, it simply *had* to be divided into 60 pieces (minute pieces, you might say), and then, because 60 is just that important, those had to be further subdivided into 60 tinier pieces (thus a "second" division).

And, as you are well aware, every last one of these pieces will come rushing at you when a crucial deadline is approaching.

Despite what you feel, though, your watch doesn't run any faster when you're under pressure or any slower during those mandatory meetings. It does turn out, though, that time is definitely relative.

5.2 Time in a Bottle—or in a Black Hole

You and the friends that you still have after the Earth-Moon-Sun demonstration decide to synchronize your watches and meet at the pizza place at exactly six o'clock, where you've promised to buy pizza and pitchers for everyone. Thanks to those Babylonians and subsequent hordes of even more time-obsessed people (remember the folks from International Earth Rotation and Reference Systems Service that we met in chapter 1?), our watches are now precision instruments, and we all know what six o'clock means, so everyone will get there at exactly the same time. Right? After all, the second hand on your watch will tick along at the exact same rate as the second hand on everyone else's watch.

Won't it?

The good news is that it's highly doubtful that any of your friends will ever do what it takes to make time tick along at a measurably different rate. The weird news is that, technically, your measurement of time depends on where you've been and how fast you've gone. What's even weirder is that time appears to cease altogether under certain, very extreme circumstances.

Yes, more extreme than that mandatory meeting.

Now if you find this a bit tough to swallow, you're in good company. Gravity-calculus-optics-motion genius Isaac Newton himself declared that time is an absolute feature of the Universe that flows independently of everything else. It would never have occurred to him to say that the cosmic heartbeat could possibly be affected by anything. Time just *is*. Unfortunately, if you had asked Newton *what* exactly time is, his response would likely have been something like, "Time? Well, that's ... um ... why everyone knows what *that* is. I don't even need to waste my energy explaining it." Although he couldn't put it into words, he was pretty confident that time is one of those Universal absolutes.

It was the scientific equivalent of a parent's "Because."

This isn't to say that people hadn't tried explaining time before Newton. Aristotle gave it a shot, describing it as some quantity related to motion and that time and motion are hopelessly entangled. Without motion to measure time against, he suggested, there's no time, and without time, there'd be no motion to observe. Case closed.

It's funny ... when philosophers say things like this, their statements get analyzed and debated for centuries. When the rest of us say things like this, particularly in a high-school essay, we get a C-minus.

Anyway, about 700 years later (that's 700 revolutions of Earth around the Sun to you, Mr. Time-Is-Measured-by-Motion), St. Augustine spent some time pondering time, mostly to have a nonsarcastic answer for people who asked questions like, "What was God doing before He created the Universe?" (The sarcastic answer—"Preparing hell for people who ask questions like that"—had already been tried, but for some bizarre reason it just alienated people.)

St. Augustine's answer was highly theological, metaphysical, and strangely Einsteinian for a guy living in the fourth century: there was no "before." Time was part of the Universal package deal, and it's wrapped up in the very material of the cosmos. Without a material Universe, there is no time. Anything outside of the Universe is necessarily outside of time, as well, which is why "eternity" to St. Augustine didn't mean "a whole boatload of successive years," or some infinite stretch that would seem to get pretty boring after a while, no matter how you were spending it. Instead it meant more like a complete sense of "present."

As crazy as that sounds, there really is something that gets to experience a complete sense of present, or eternity, or timelessness. That something is light. One of the other bizarre things that Einstein discovered while performing all his Universe-bending thought experiments is that the faster you go, the slower your clock ticks. It turns out that the Universal speed limit is the speed of light, or 186,000 miles per second. The closer you get to that speed, the more time slows down.

Once again, do not use this information when your alarm doesn't go off Monday morning. You'll have to zip to work at a hefty percentage of the speed of light (at least the red lights will look green, though), and once you come screeching into the parking lot, you'll find that your boss's clock still shows that you're late.

Still, it's been shown that things happen more slowly than expected for really fast-moving particles. It's as though they're more casual about getting things done when they're traveling at high speeds. There are experiments with bizarre particles called muons, which sound like something made out of tiny cats. Apparently, if muons are left sitting by themselves, they will spontaneously change their identities within about 3 millionths of a second. If you get these little suckers moving 99.99% the speed of light by running them into a particle accelerator, they'll stick around as muons for a good hundred times longer. Their internal clocks still think they're falling apart on schedule, but we outside observers get to watch them live much longer when we make them go faster.

Quite the opposite is true for humans, though. For some reason, the faster we go, the shorter our lives are, at least when driving. Perhaps a new anti-speeding public service campaign should be, "Stop trying to be a muon." But I suspect only you and a handful of particle physicists would get it.

At any rate, decades of experiments have borne out this strange notion that the steady "absolute" of time is nothing like steady or absolute, except in the case of light. At light speed, time simply stops. We may witness a beam of light trek across the solar system, taking hours to get from the Sun to Pluto, but the light itself does not experience the passage of any time at all. None. It doesn't age a bit along its journey.

Finally—the solution to jet lag!

Tons of science fiction writers have imagined ways to hitch a ride on a light beam, and not just to fight jet lag. A trip to another planet, or even another *galaxy*, would be instantaneous. The trip back home would be, from your perspective, instantaneous.

Unfortunately you won't be able to give out all those great extragalactic souvenirs to your friends and family ("My pal went to Pal 4, and all I got was this koozie!"). Although your trip seemed to go by in a flash, everything on Earth will have aged millions or even billions of years.

Another severe drawback is that, as far as we know, only one thing is capable of traveling the speed of light—you guessed it, light. Everything else has to settle for some lesser speed. We can't even get those muons to speed up that last tiny percent of a percent. The faster things go, the harder it gets to make them go faster, until it literally becomes infinitely hard to make them go the speed of light. Either you're already moving that fast or you never will.

It's law enforcement at its finest.

If that doesn't make your brain hurt, this should. It's not just moving that makes clocks tick at different rates. St. Augustine now seems positively prophetic in his notion that time really does seem to be woven into the material of the Universe. That's why the huge elastic sheet with the bowling balls and confused

Yes, I'm sure! The GPS says the restaurant is 200 feet on the right!

ants in chapter 4 was called the fabric of space*time*, not just the fabric of space. It's not just the spatial gridlines that are stretched out by mass. The time ones are, as well.

What this means is that those cesium-133 atoms—you remember, those hyperactive little things that oscillate 9,192,631,770 times a second—aboard an airplane jetting around Earth won't appear to keep the same beat as the cesium-133 atoms sitting on the ground, snuggled deeper in the dip closer to our planetary bowling ball. The farther away you get from a fabric-stretching mass, the *faster* your time will tick.

Technically—*technically*—your head is aging a teensy-weensy bit faster than your feet, and tall people age a teensy-weensy bit faster than short people. So maybe *that's* why everything seems to take so long when you're a kid. Time really does go more slowly for them!

Seriously, though, how much of a difference could any of this possibly make in your life? Plenty if you rely on anything at all that uses the Global Positioning System, or GPS. In their orbits 20,000 miles above Earth's surface, GPS satellites have atomic clocks that tick at a measurably different rate from their Earth-bound counterparts. Their zippy motion makes the satellites' clocks tick more slowly, but this effect is overshadowed by the gravitational speeding-up of their clocks that results from their greater distance from Earth.

In just a few minutes, without taking the stretching of time gridlines into account (or, more precisely, without considering general relativity), a calibrated GPS system would get your position wrong by several yards. After a day, it would think you were miles away from where you really were. Without understanding the way time and space and motion and matter are all linked together, you could just forget about navigating through an unfamiliar city with your car's spiffy new system. But true disasters would occur if we failed to take this into account in flight navigation. This sort of accumulating error just won't do if you want to

land a plane safely, which is why these and other satellites are closely monitored and often recalibrated.

It might be bizarre and unbelievable, but it's just the way the Universal clocks operate.

Speaking of bizarre and unbelievable, the situation gets really peculiar inside the dips caused by the most insanely compressed objects. Formerly known as "gravitationally completely collapsed stars," black holes (which I'm sure we all agree is a far better name) are places where gravity is so intensely strong that nothing—not even light—can escape. In the stretched grid picture, these are bottomless pits. No matter how hard you attempt to roll a ball up the sides, it'll never make it out. And if a ball rolls down into the endless pit, the "time" gridlines will get ever farther apart, stretched until, at the edge of the pit itself, the distance between them becomes infinite. If you were to watch something equipped with those oscillating cesium-133 atoms head toward a black hole, eventually you'd see the oscillations stop completely.

That is, from an outsider's perspective, time just stops inside a black hole.

What's really bizarre is that the cesium atoms, except for the unfortunate fact that they'd be shredded proton from proton by the insane tidal forces as they plummeted into the cosmic abyss, wouldn't notice a thing happening to time. If they could manage somehow to stay intact, they wouldn't stop what they were doing. We just wouldn't get to watch it. In this sense, black holes are the Universal censors, and even though they won't let us know what's going on inside those bottomless pits, we can be pretty sure that it's nothing like what's going on out here.

There is speculation, though, that both space and time cease to exist in the very hearts of black holes. Why they should *stop* existing is as puzzling as why they exist in the first place, and unfortunately all of the physics currently at our disposal can't plumb the heart of a black hole.

5.3 A One-Way Ticket to Disorder

If you think of space-time as a rubbery sheet of gridlines with bowling balls distorting its surface and occasional bottomless pits, one of the things you might wonder is why you can go back and forth on a certain type of gridline, but you're stuck on a one-way street when it comes to time.

Think about it. If you want to move toward your refrigerator, you can. If you want to move away from it, you can do that as well. We really don't appear to have any problem moving within the dimensions of space. Highways go north and south, after all. Frustratingly, however, even though time is apparently connected to the spatial packaging of the Universe, we can't seem to choose our direction in time the way we can choose whether to go north or south on the highway.

When it comes to time, we're stuck on a one-way street. It's a harsh and unforgiving reality. There's probably not a thinking person alive who hasn't at some point wished to go back in time—even just for a minute or two—for a do-over.

But why is it just one way?

This is probably the best place of all just to say, "Because."

I know. The kids never go for that answer, either. But to figure out why time behaves the way it does would require a pretty good idea of what time actually is. While we're fantastic at keeping time, we're pretty pathetic at knowing what it is. This is a problem unique to time-keepers. A zoo-keeper is quite aware of what it is that he's keeping, but if you ask a time-keeper what exactly is being "kept," you probably won't get much of an answer. In the timeless words of St. Augustine, "If no one asks me, I know what time is. If I wish to explain it to him who asks, I do not know."

As far as we can tell, though, time does exist and, as Groucho Marx (or maybe it was Isaac Newton) pointed out, it does fly like an arrow. That is, it seems to have a particular direction. Discovering the reason behind this directionality might be as simple as following a game of gossip.

Come on ... don't pretend you didn't play this in elementary school. Someone would come up with a simple sentence like, "I enjoy playing in the backyard, both night and day." She'd whisper it to the kid next to her, who'd whisper what he thought he heard to the kid next to him, and by the time it got around the room, the last kid would look really puzzled and say, "No flailing of pecan shards on the right of way?"

And everyone would laugh because the sentence got completely mangled as it went down the line.

This is precisely the sort of behavior we see in the Universe: a tendency toward disorder. In science, that tendency isn't called "gossip," though. It's called entropy. Basically it tells us that the Universe as a whole—and any system, for that matter—is slowly becoming less and less organized. In fact it apparently *must* progress from order to disorder.

My entire house is a testament to the reality of entropy, and I can't imagine anything more bizarre than to enter my son's room and find out that some kind of entropy reversal had taken place.

The reason the game of gossip is so fun is that there is only one way to keep the original sentence, but a zillion ways to mess it up while keeping something like the same rhythm and tone. From one person to the next, it becomes slightly disordered. Perhaps instead of saying "night and day," he says, "right away." Then the next person messes it up a bit more, and on it goes. It would be pretty miraculous if the last person in line managed to reconstruct the original sentence, but it wouldn't be impossible. It's just highly unlikely. In fact, if there is only one way for it to be right, but a zillion equally likely ways of messing it up, the chances are one in a zillion that it'll wind up being the original sentence after the game is over.

And the worst part is that it wouldn't be funny at all.

Whoah! Now THAT'S creepy!

So perhaps the Universe is just trying to see what kind of crazy final sentence will come of all this, because it, too, is slowly becoming more and more disordered. It doesn't have much choice in the matter. Just like in the game of gossip, if the Universe goes from one state to the next, the odds are a zillion to one that it'll be more organized than before. And so our brains have a perception of time that follows this direction of increasing disorder. We experience the Universe as it progresses from one state to the next, and this forces the arrow of time to point in the same direction as the arrow of disorder.

So what's becoming disordered? As far as we can tell, the Universe is currently expanding, and energy is slowly getting squeezed out of stars. Hydrogen is being used up, and eventually the fuel to make stars shine will be gone or so spread out that it can't come together to make new stars. In billions upon billions of years, if cosmologists are anything like correct, there won't be any solar systems or stars or life or anything, really, except for some stray subatomic particles and a whole lot of low-energy radio waves.

The epitome of disorder.

It's a pretty depressing prospect for our spectacular Universe, really.

But back to that thing called time. If you've been paying attention, you'll have no doubt noticed that the current cosmological concept of time is actually the same frustrating definition that Aristotle trotted out over 2,000 years ago, but in more sophisticated packaging. These days, instead of saying that time and motion are somehow inextricably linked together, we say that time and entropy are somehow inextricably linked together. Neither the past (whatever that means) nor the present (whatever that means) definition gives us a scrap of insight into what time is. But at least our current definition provides for some fascinating possibilities.

One of those possibilities is the slim—*very* slim, I should mention—chance that the Universe might stop expanding and be brought back together by the mutual attraction of everything in it. Should that happen, there's a school of thought that suggests that things would start happening "in reverse." You'd remember tomorrow's exam. You might even "grow" from old age to infancy, just like the title character in the movie *The Curious Case of Benjamin Button*. But it wouldn't be "curious" because everyone would be doing it.

On the other hand, apparently *nobody* would be doing it, because another school of thought insists that intelligent beings like you and me can't exist in a contracting Universe.

It's enough to make your brain go on strike.

If you want it to really rebel against you, then consider this: some scholars have "proven" that time doesn't even exist.

And it just gets uglier from there.

5.4 Time: The Grand Illusion?

So, in the first chapter, you found out that the night sky isn't actually dark at all. In the second chapter, you learned that light isn't a wave or a particle, but somehow it's both. In the third chapter, you found out that some of your favorite scientific truths were lies. And in the fourth chapter, you found that the force of gravity isn't actually a force. It should come as no surprise, then, that the thing that entire societies are built around might not even exist.

And, no, I'm not talking about the stock market.

It's tempting to blame it all on Einstein, who said once that "the distinction between past, present and future is only a stubbornly persistent illusion." How could something so real be unreal? We measure it, for heaven's sake! How can you measure something that's not there? And as for past, present, and future. . . . Obviously you finished reading the previous sentence at some time in the past, which was definitely not the same as the time you're reading this.

Or this.

Or even this.

Philosophers wrestle with words and ideas all day long, but it's hard to take their proofs very seriously sometimes, especially when one philosopher proves the exact opposite of what another philosopher proves. Philosophers even have issues with their own existence, which is why René Descartes had to "prove" his own reality by saying "I think, therefore I am."

When you have to prove your own existence, your day just can't get much weirder.

Mathematicians and physicists, on the other hand, seem to have a knack for knowing what's real and unreal. You can measure things about muons. You can mathematically prove that two triangles are similar. And no matter what you might think of the reality of your own existence, you can duplicate these measurements and proofs for yourself and get the same answers.

So when philosophers say things like "time doesn't exist," we can usually go on with our day, content in the knowledge that another philosopher disagrees. But if a brilliant mathematician starts tinkering with Einstein's relativity theory and proclaims that time doesn't exist, you have to sit up and take notice.

That's just what Kurt Gödel—Einstein's close friend and colleague at Princeton's Institute for Advanced Study—managed to do. Proving that time didn't exist, though, only followed his mathematical proof that the set of mathematical proofs was necessarily incomplete, something that makes absolutely no sense to us nonmathematicians but which really threw the mathematical world for a loop in 1931. Once he'd completely shattered the foundation of one field, Gödel was free to destroy the foundations of other fields, namely physics, as only someone with a reputation like his could.

Einstein himself never could stand mathematics. Sure, he could calculate just fine, but he was convinced that real truths were not found in symbols, but in thought. Linking time to space and demonstrating that measurements of past, present, and future are "relative" was only later put into mathematical form by his old math teacher, who was sharp, but nothing quite like Gödel. When a mind like Gödel's gets hold of a problem, it takes no prisoners. He didn't just formalize Einstein's ideas. He saw things in relativity that nobody else—even Einstein—could see.

If Einstein were right, Gödel demonstrated, then there were real, physically possible universes where time travel could happen. And if you could physically travel to the "past"—even in a hypothetical universe—then the time didn't really pass, did it? For instance, if you want to travel to New York and then to Boston, that doesn't mean you can never return to New York, right? You can always just turn around and head back to New York.

But let's say that you have some concept of the Universe where Boston *must* happen after New York. That Boston is somehow a future event and New York is a past event; that somehow New York stops existing once you get to Boston. Just the fact that you can return to New York means your whole concept of future and past is wrong. In other words, your idea of time—the property of your Universe that forces Boston to happen after New York—*must* be incorrect.

In the Universe wrought by Gödel out of Einstein's relativity equations, events don't have to flow into the past. Future and present and past don't exist. You can travel around from one "time" to the other just as you can get from city to city and back again.

Ouch.

If this makes precious little sense to you, don't worry. Most mathematicians and physicists and philosophers were either so confused or so taken aback by Gödel's proof that they did what any of us would do: they ignored it.

The fact that Gödel was eccentric to the point of paranoia also made his idea easy to dismiss. Why listen to the ramblings of an insane man, after all?

Famed physicist Stephen Hawking wasn't afraid to bite, though. He tackled the problem of so-called "closed time-like paths" (English translation: real ways to travel in time) head on and concluded that the Universe just doesn't like people messing around with time, so it doesn't let us. So there. The Chronology Protection Conjecture was, thus, invented (okay, there was a bit more thought and mathematical rigor involved). Sounding a bit like some government committee, the CPC basically keeps the arrow of time pointing in the right direction.

I can see the public service announcements now. "The CPC: Working to Keep Yesterday and Today Safe from Tomorrow."

Sure, there might be some slowdowns here and there—near gravitating objects or at high speeds—but nobody's going to just turn around and head backward, no matter what the equations allow. The CPC essentially forces time both to exist and to flow in a certain direction. Furthermore, it answers the pesky question of why time-tourists from the extreme future aren't perpetually bothering us, and we don't have to humbly admit that perhaps they just don't find us past-dwellers all that interesting.

One thing is certain, though. You'll never look at your watch the same way again.

 **SMALL WONDER**

So What Causes Daylight Saving Time to Happen?

Politicians.

 **SMALL WONDER**

Why Are There Seven Days in a Week?

The origin of our seven-day week goes way back to ancient

THE
CPC
WORKING
to keep yesterday and today safe
from tomorrow
Call 666-666-6666
TO REPORT TEMPORAL BREACH

times. There are seven objects in our sky that have fairly distinct motions. The Sun is one of them—the most obvious, in fact—so it gets its own day: Sunday. The Moon is the next most obvious thing, and even though it appears to rise and set with the stars, its motion is ever so slightly different. Remember the interpretive dance you did with your friends and family? Better give it a day as well: Moon-day, or Monday. The rest of the days are named after the planets you can see with the naked eye. Like the Moon, they have motions that are subtly different from the rising and setting stars. In fact, if you're really patient (years of patience, in some cases) everything in our solar system will appear to move against the background of distant stars. Before telescopes, though, the only objects that we could see misbehaving like this were Mercury, Venus, Mars, Jupiter, and Saturn. In English, only Saturn's day is recognizable, but if you look at, say, Spanish, you'll see that Mercury has its own day (Miercoles—Wednesday), Venus has its own day (Viernes—Friday), Mars has its own day (Martes—Tuesday), and Jupiter has its own day (Jueves—Thursday). (I know it doesn't sound much like Jupiter, but that's because it's named after Jove, which is Jupiter's nickname, by Jove.)

Imagine if the ancients could have seen Uranus or Neptune! We might have a nine-day week with Uranaday and Neptunaday rounding it out. Then we could have three-day weekends every week. The down side is that we'd have six consecutive work days, but at least we'd be able to take more weekend trips.

Monday would still be painful, though. On the other hand, since time doesn't actually exist, what difference will it make if you sleep in those mornings?

SMALL WONDER

What Would Happen if You Fell into a Black Hole?

It's a tempting thought, especially for science fiction writers. If black holes can distort time and space so much, perhaps there's a way for you to travel through them and find yourself in a different place and a different time. Fueling the fires of these fantasies are mathematical models for things called wormholes, which do seem to transport matter and energy from the black hole in one part of spacetime to a white hole in a completely different part of spacetime.

There is one minor problem, though. If you fall into a black hole, you will die.

"But what about . . ."

No, really. Trust me. You will die.

If you fall feet-first into a "normal" black hole (one left by a collapsed star and that has over three times the mass of the Sun compressed into a dot like this: •), the gravitational tug on your feet will be so much greater than the gravitational tug on your head that you'll be stretched out like a bit of cosmic taffy. As you got closer to it, your very atoms would be ripped to shreds.

It wouldn't be a pretty sight.

Fortunately for those of us watching you, we wouldn't have to witness any of that because your time will appear to go slower and slower as you get closer and closer. If, on the other hand, you find an actual wormhole—a highly unlikely prospect, by the way, because it's just a mathematical solution to some equations and probably not a physical reality—you'll still die. Just your presence in the thing will cause it to collapse in on you. Even if it didn't collapse in on you and kill you, all the high-energy gamma rays in the vicinity of the thing would finish you off.

No matter what, you're not going to like it. So it's better just to stick to the here and now, even if those don't actually exist.

CHAPTER 6

Home

*There is nothing like
staying at home for real comfort.*
—Jane Austen

I magine summer. And kids. Really, really bored kids complaining that there's nothing interesting to do. Compounding the problem is the lack of money to do anything. Fortunately, camping doesn't cost much. So everyone eagerly loads up the tent and marshmallows and other necessities of life and you head to the campground so that, at the very least, you're not staring at the same boring walls for a long weekend.

Within approximately 48 minutes of arrival, there will already be complaints about how hot it is. After just a day, the complaints will be in full bloom. Sleeping bags aren't as comfy as the bed. Breakfast cooked on a camp stove isn't as good as it is at home. The bathroom's too far away.

No . . . really. It's *way* too far away.

By the end of the weekend, everyone is starving for home. And, as L. Frank Baum pointed out in his classic *The Wizard of Oz*, it wouldn't matter if you spent the weekend singing and dancing with sentient scarecrows in the magical land over the rainbow. You'd still long for home.

But what's so special about that place, anyway? Plenty, as it turns out.

6.1 Goldilocks and the Three Planets

Just like when you camp at your local state park, you don't have to go too far, cosmically speaking, to appreciate your home. And as hot as the campground might be in the summertime, it'll never rival the conditions on our nearest planetary neighbor, Venus. Orbiting every 225 Earth days at a relatively earthy distance of 67 million miles from the Sun, Venus has traditionally been known as our sister planet.

Perhaps Evil Twin would be more appropriate.

It's hard to escape the irony behind all the nicknames of our solar system's second planet. One of the more striking objects to show up in our evening or morning sky, it's often called the Evening Star or the Morning Star, despite the fact that it's not a star at all. Even more ironic is that the Romans named it for the goddess of love, which makes it the poster child (or poster planet, as the case may be) for jilted lovers and cynics. "Sure, it looks pretty from the outside, luminous and inviting, but once you actually get there, it's a horrible, stifling, crushing, hellish place."

Astronomically, we weren't exactly doing a great job of pinning down Venus's personality, either. You see, most visible light reflects straight off the cloud tops of Venus, so it always looks essentially bright beige. This makes it quite a challenge to figure out even the simplest things about Venus, like its rotation rate. For the longest time, people were torn between the notion that Venus had a day identical to ours and that Venus was tidally locked to the Sun, one side in perpetual daylight and the other in perpetual night. It wasn't until we took a look at it in microwaves and radio waves, types of light that are able to penetrate the clouds, that we got some information about what things are like there. Then we plunged a few spacecraft into its atmosphere to find out the whole truth.

Unfortunately, it wasn't anything like what we had hoped.

The problem was and still is those darned clouds. We humans

have certain expectations about places with clouds. They should be wet and swampy, for one thing. (If you ever read the relatively obscure C.S. Lewis book *Perelandra*, you'll be swept up in his visions of floating continents and a vast ocean of a planet.) Even people who were trying to look at Venus scientifically were pretty convinced that it had to be essentially a warmer, wetter version of Earth. After all, its size and mass and distance from the Sun were almost identical to Earth's, so it should be pretty similar. A few calculations based on its distance, reflectivity, and amount of sunlight told us it should be a bit hotter than Earth, but not painfully so.

Boy, were *we* in for a surprise!

The first thing we got good information on was its rotation rate. It's not actually tidally locked to the Sun but instead rotates once on its axis every 243 days. Furthermore, it does this "backwards," so that the Sun rises in what we'd call the west and sets in the east. Not that any of that matters because you'd never actually *see* the Sun from the surface of Venus. You can thank the thick Venusian clouds for that.

Well, that and you'd be dead.

You might think that its thick carbon dioxide atmosphere wouldn't be all that bad, really. A few hardy plants here and there to convert it to oxygen and—poof!—we've got a breathable atmosphere, right? Except for the fact that the atmospheric pressure is about like being half a mile under the surface of the sea, where you'd feel 1,300 pounds pushing in on every square inch of your body. If that weren't bad enough, the temperatures aren't just a little bit warmer than Earth's. They're beyond broiler-oven hot at nearly 900 degrees Fahrenheit. All over, day and night, with no relief. It's so hot, in fact, that the rocks on its surface are thought to be softened to the consistency of warm lead.

Not that you'd want to scrape your fingernail along them or anything to find out for sure.

Scattered along the surface of Venus are several former Soviet spacecraft. Former in the sense that the Soviet Union no longer exists, but also in the sense that the probes have been reduced to corroded blobs at this point. From the 1960s through the 1980s, the Soviet Union sent a veritable armada of spacecraft to study Venus. Several of these actually landed on the surface, snapped some panoramic pictures, took readings on the wind speed, atmospheric conditions, and then promptly died after about 90 minutes. On their way to a slow, crushing death on the ground, they passed through the Venusian upper clouds, which are composed of sulfuric acid. Acid rain (actual acid, not just slightly acidic water like we get here) falls from these high clouds, but it never manages to hit the ground.

Even during peak mosquito season, your local campground is definitely a better vacation spot than Venus.

So how did it get to be so nasty there? Even the daytime mercury on Mercury, orbiting a mere 36 million miles from the Sun, barely tops 900 degrees, so how did a planet nearly twice Mercury's distance from the Sun become our solar system's pressure cooker?

It's an amazingly complicated question, one that planetary climatologists have wrestled with for decades, partially because Venus is tough to read (nothing survives long on its surface, and peering through its thick atmosphere is not exactly trivial) and partially because it has a 4.5 billion–year head start on all our observations. But the essentials seem to boil down to a few key problems.

One major problem with Venus is the lack of liquid water. Oh, sure, there were probably oceans of it to start with, but being a good 26 million miles closer to the Sun than Earth, Venus was just a bit too toasty to maintain them for more than a billion years. So they began to evaporate.

Water vapor, though, is notoriously good at holding in heat. Think about the crystal-clear, low-humidity days here on your home planet that are perfect for rubbing balloons on the heads of kids and then sticking them to the walls. The balloons, that is. During the day, the temperature might work its way to a balmy 75 degrees Fahrenheit or so, but at night, the temperatures can dip below freezing. On the other hand, on the days that you wouldn't be surprised to see fish swimming by your windows, the difference between daytime and nighttime temperatures is sometimes nonexistent. Water vapor is practically a blanket for us.

Unfortunately, if your planet is a bit too hot to begin with, you don't really need an additional blanket. All that does is help make things hotter, which causes more of the oceans to evaporate, which only helps hold in more heat, which causes more of the oceans to evaporate, which . . . You can probably see where this is going. It's a snowballing effect or, in the lingo of planetary climatologists, it's a runaway greenhouse effect.

The greenhouse effect is quite possibly one of the most politically charged topics in all of science these days, but it's a pretty straightforward concept, scientifically speaking. Certain substances are fantastic at allowing visible sunlight through. These same substances are sometimes terrible at allowing infrared light through, however. Once the sunshine hits the ground and heats it up, the ground will emit infrared light. This infrared light then tries in vain to get back out into space, but instead it gets stuck inside the atmosphere like a pixilated square of a ball in the ancient video game of Breakout. Water vapor helps hold that energy in, as does carbon dioxide.

Those two substances are pretty much all the original Venusian blanket was woven from, so the poor thing got really hot really fast. As temperatures rose, carbon dioxide gas was even leached from the very rocks, making the atmosphere an even *more* efficient trapper of heat. All the while, the poor water molecules that were now part of the clouds rather than the deep blue seas became the victims of Venus's second big problem: no magnetic field.

On the list of things you attribute to your continued survival—

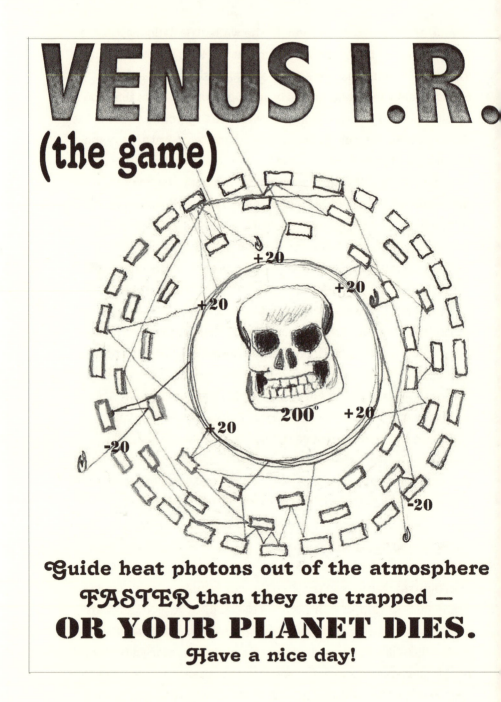

you know, air, water, food, etc.—Earth's global magnetic field is probably around number 38,162. After you finish reading the next few paragraphs, it'll most likely be in your top ten, because our magnetic field does more than just give compass makers a job and help prepare scouts for those treks into the wilderness. It does a darned fine job of keeping us from becoming victims of stormy space weather.

As you've probably noticed, there's a massive hydrogen fusion reactor just 93,000,000 miles away. It's not just quietly killing off electron evil twins and making life-giving sunshine in the process. It's also creating a chaotic, roiling cauldron of ionized gases and turbulent magnetic fields that have a nasty tendency to burp out colossal blobs of charged particles that occasionally come racing toward Earth. Standing in front of a wave of fast-moving ("fast" = hundreds of miles per *second*) electrons is Not a Good Idea ("Not a Good Idea" = potentially fatal). Less so is being exposed to a wave of fast-moving protons.

Remarkably, we're actually still alive. I say "remarkably" because billion-ton blobs of this stuff, along with high-energy light, regularly come spewing off the face of the Sun. (Want to know what the Sun's spitting at us right now? Check out www.spaceweather.com.) One thing that helps us weather these space storms is Earth's magnetic field, which does a dandy job of deflecting these charged particles and funneling energy toward our poles. If you live in a far northern or southern part of our planet, you even get to witness the spectacular light show that is produced when these particles meet our magnetic field and are sent to the upper atmosphere. It's called the aurora and is one of the more amazing sights you will ever witness from planet Earth. Be sure to thank our magnetic field and atmosphere next time you catch one. (And if you've never managed to catch one, put "viewing the aurora" on your lifetime to-do list, sandwiched in between finishing this book and dipping your toes in every ocean.)

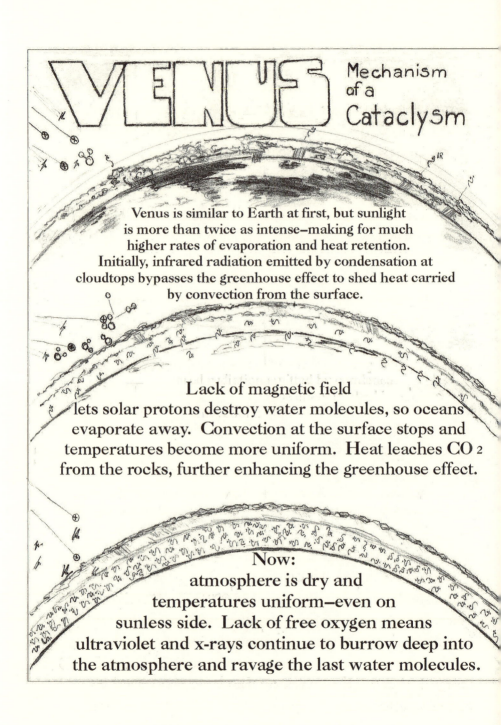

VENUS
Mechanism of a Cataclysm

Venus is similar to Earth at first, but sunlight
is more than twice as intense—making for much
higher rates of evaporation and heat retention.
Initially, infrared radiation emitted by condensation at
cloudtops bypasses the greenhouse effect to shed heat carried
by convection from the surface.

Lack of magnetic field
lets solar protons destroy water molecules, so oceans
evaporate away. Convection at the surface stops and
temperatures become more uniform. Heat leaches CO_2
from the rocks, further enhancing the greenhouse effect.

Now:
atmosphere is dry and
temperatures uniform—even on
sunless side. Lack of free oxygen means
ultraviolet and x-rays continue to burrow deep into
the atmosphere and ravage the last water molecules.

Okay. That takes care of the zippy charged particles from the Sun, but there's still the high-energy light to contend with. On Earth, the primary thing keeping us safe from that threat is our ozone layer, a layer of oxygen molecules high in the atmosphere. Thanks to photosynthesizing plants, we have a pretty decent amount of oxygen to make a good ozone layer, which in turn prevents the high-energy light from burrowing into the atmosphere and destroying molecules that we care about, like water.

Poor Venus has neither a magnetic field nor an ozone layer. All that solar nastiness just slams mercilessly into its upper atmosphere all the time, ripping apart hapless water molecules. Over its lifespan, Venus has not only lost all its surface water oceans to evaporation, but it has also lost its atmospheric water vapor to solar destruction. These days Venus is bone-dry, even in its clouds, while its temperature and surface pressure are hellish.

Not exactly the kind of place you want to spend much time, is it?

It's really sobering to note that Venus really isn't all that different from Earth, cosmically speaking. It's practically the exact same size and distance from the Sun, and probably experienced very much the same early childhood. But a number of relatively small differences conspired to make it a lifeless wasteland, whereas here on Earth, it seems the entire surface is thriving with some kind of life, even if it is occasionally bizarre and alien to us land-dwellers.

In the words of Goldilocks, our planet is just right.

So if Venus is too hot, and we're just right, that makes Mars a cold bowl of porridge. About 50 million miles farther from the Sun than Earth, Mars has been the subject of science fiction for as long as the genre has existed, and whenever people talk about aliens, they almost unanimously call them Martians. Heck, it's even been a candy bar.

As recently as a century ago, professional astronomers like Percival Lowell even thought there was sufficient evidence that

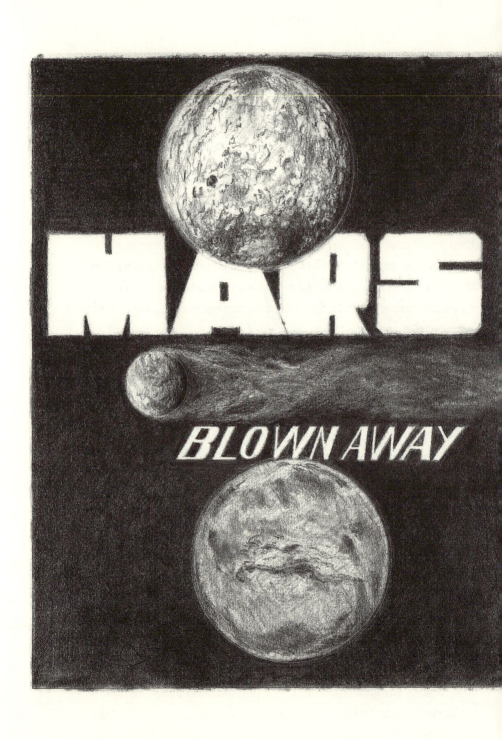

intelligent (and desperate) creatures inhabited the red planet. And these days, although we're pretty sure there aren't big critters calling Mars home, the news outlets pounce on every hint that astronomers have discovered some kind of microbial life there.

Unfortunately, Mars is not exactly an ideal home-away-from-home, either. However, it might prove to be a good fixer-upper opportunity for us some day. So many things about it seem just right: its day is almost the same as ours; it actually has seasonal variations; and it has water, albeit almost exclusively frozen. For the time being, though, it's as inhospitable as Venus. Just in a different way.

The main problems with Mars are (1) that it's a pretty scrawny planet, and (2) that it's farther from the Sun than we are. With about half the diameter of Earth and about a tenth the mass, Mars would pull on you with just under 40% the strength of Earth's gravity. Unfortunately, this means that it pulls on, say, an atmospheric molecule with just under 40% the strength of Earth's gravity, too.

Now you might not really appreciate the fact that air molecules are actually made out of stuff. It's easy to miss, since it doesn't exactly get in our way most of the time. It turns out, though, that a typical bedroom holds about 100 pounds of air. On Mars, that exact same box of air would weigh a bit less than 40 pounds. But the air in Martian boxes isn't nearly as thick as it is in Earthly boxes because that feeble gravitational tug simply can't hold much of an atmosphere to the planet. In reality, a bedroom's worth of Martian air would weigh only a few ounces.

Sure, back in its heyday, Mars had a pretty thick and respectable atmosphere, replenished by plenty of volcanic outgassing (Martian volcanoes make even Mt. Everest look puny—too bad they're all long extinct). It was even gushing with liquid water and, apparently, had a magnetic field to protect the atmospheric molecules. But it just wasn't enough. Because it's so small, Mars cooled relatively quickly, solidifying an ever-thicker crust. Volcanoes and other things that can help keep an atmosphere replenished simply died away. Along the way its planetwide magnetic field disappeared, too, allowing the Sun's energy to rip away the atmosphere. The weak gravity was simply no match for it. Before Mars knew it, its warm, wet, thick atmospheric blanket had either been kicked into interplanetary space by the solar wind or frozen into the ground.

Unlike Venus, Mars still has oceans of water, but the water's all frozen underground or trapped in the Martian polar ice caps. Carbon dioxide fog occasionally fills the deeper valleys and wispy carbon dioxide clouds scurry around the surface. Despite the fact that virtually all its atmosphere is an efficient greenhouse gas, there just isn't enough of it to trap the Sun's energy. Consequently, Mars is now a frozen wasteland, with an average temperature of less than −40 degrees Fahrenheit. That might sound livable to you folks who regularly deal with those sorts of temperatures in Alaska, but the cold temperatures would be the least of your problems. On Mars, the atmospheric pressure is so low—about 1% of that on Earth and about 1% of 1% of that on Venus—that your ears would immediately and painfully pop, any air in your lungs would go whooshing out of your system, and your blood would boil within seconds of being exposed.

This sort of makes camping on Mars a tough sell, too.

Again, though, if Mars had been just a little bigger or just a little closer to the Sun, it would have been "just right." As it is, our solar system seems to sport only one place that's not too hot, not too cold, not too big, not too small, not too hard, and not too soft for life to thrive: our home.

6.2 Goldilocks and the Three Stars

It's not all Earth's doing, of course. The star that we orbit deserves a huge amount of credit for being "just right," too.

Remember Gliese 581 from chapter 1 and its attendant planets that might be tidally locked to it? This stellar lightweight
has less than a third the mass of our Sun and pumps out only
1% as much energy, most of which is invisible to the human eye
because it's infrared. It's also far cooler than our star—a chilly
5,300 degrees versus 11,000 degrees Fahrenheit. If you plopped
Earth 93,000,000 miles away from this puny reddish thing, we'd
be as much a frozen wasteland as Mars. And although the tidally
locked planets make for an interestingly nonexistent day-night
cycle, that sort of behavior is not exactly the ideal arrangement
for life to gain a foothold. Being that close to the star—even a cool,
dim one—puts the planet's entire atmosphere at risk for being
obliterated by violent solar flares. If it manages to survive those,
the endless, cool night on the dark side could cause the atmosphere to hit its dew point.

Then there's the whole issue of water. Somehow it's got to get
to those planets, and astronomers generally agree that it's part
of the package deal that involves the original spinning pancake of
stuff that the solar system is made from. The problem is that the
closer you are to the central star, the less likely water is to survive being in your pre-planet pancake neighborhood.

These potential problems aren't carved in stone, mind you.
Plenty of astronomers have worked out ways around these
issues, but they sure make us more pessimistic about the prospect of finding some kind of intelligent creatures around a low-
mass star.

So what about a higher mass star? One that's hotter and
that the planets don't have to hug quite so closely? Well, these
guys have their share of problems as well. For one, they don't
live very long, which seems completely backwards because
they're ... well ... massive. Consequently they have huge fuel
tanks. Contrary to what you'd expect, it's the low-mass stars
without much fuel to maintain hydrogen fusion that last far
longer than something like the Sun; high-mass stars with oodles

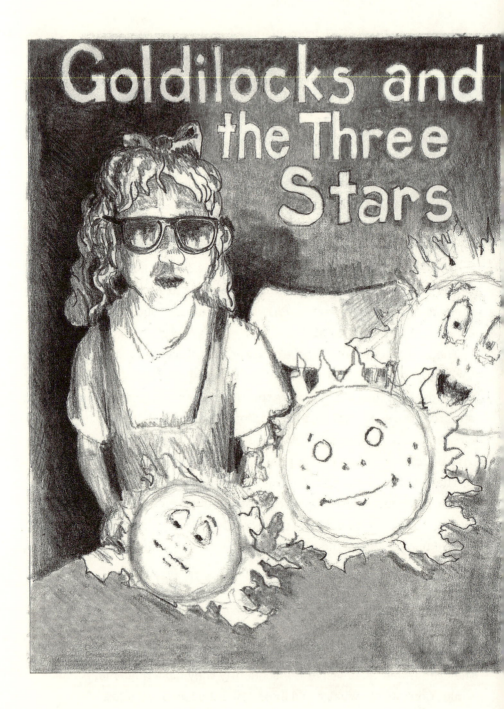

of hydrogen fuel use it up in a cosmic heartbeat. High-mass stars are cosmic gas guzzlers, you see, while lower-mass stars are more like efficient little fuel sippers, burning their tiny little hydrogen reserves a thousand or even ten thousand times more slowly than the Sun. The enormously massive stars can use theirs at a rate that's thousands or even millions of times greater than the Sun's.

This cuts their lives short by quite a bit. Even a star with just 50% more mass than the Sun will be gone before a potential life-sustaining planet gets its act together. First there's the solar system to clean up. As the stuff to make planets is busily pulling itself together, there's an enormous amount of jostling and bumping and colliding. In fact, it can take a good billion years or so just to clean up the majority of the debris that can slam into a growing planet and wipe out any bits of life that might have gotten started.

Even Earth—4.6 billion years in the making—is still a potential target for killer asteroids and comets. Just think "dinosaurs." Those guys were destroyed practically yesterday (about 65 million years ago) in the grand scheme of things. Things were far worse a few billion years ago in the turbulent childhood of our solar system.

A growing planet around a star much more massive than ours barely has time after an era of massive collisions to consider the possibilities ('Hmm . . . I think maybe I'll start work on a microbe this giga-year . . .') before its star rudely pops out of existence.

As it turns out, the only stars that are truly excellent candidates for hosting life-giving planets are, as you might have guessed by now, ones not too hot and not too cold. In other words, not too much different from the Sun. These medium-mass, medium-temperature stars live long enough to allow a solar system to become a slightly less dangerous place to raise a nice planet family, and their energy output is great enough that a planet doesn't have to practically touch them to have the right conditions for life.

Once again, our solar system proves to be "just right." Goldilocks would just be ecstatic.

6.3 Goldilocks in Suburbia, Traffic, and Time

Before she can get too comfy, though, Goldilocks needs to know that it's not enough simply to have the right planet at the right distance from the right star. The star itself has got to be in the right neighborhood and moving the right way, too. (Have you ever noticed that when you see a word like "right" often enough, it starts looking . . . well . . . not right?) Plop the Sun near the center of our Galaxy—you remember, the place where the night sky is blazingly bright from the overcrowded conditions—and the lethal radiation of frequently exploding stars would put a quick end to any life trying to make a go of it.

Heck, even being around a star that didn't have a nearly circular orbit out here in the Galactic suburbs would be dangerous. To appreciate this, go to an ice-skating rink. At first, be a considerate ice-skater, gliding slowly around the perimeter at roughly the same speed as everyone else. See what sorts of reactions you get. None, right? Nobody's cussing you out; people aren't swerving out of your way. Everyone's well behaved.

Now be on the lookout for someone a bit more obnoxious. Someone who's skating in and out of traffic, zipping in front of people. See what kinds of reactions that person gets. You might not witness any collisions, but people are definitely being affected.

In the case of the Sun, the more in line it is with the local Galactic traffic pattern, the less likely it is that our solar system will be gravitationally affected by other stars. Like the ice-skaters, the stars don't have to actually collide for there to be noticeable consequences. (Interstellar distances are far too great for collisions, anyway.) Don't worry—Earth's not in any danger of being pulled out of orbit. It's a far more subtle effect involving solar system objects far beyond the orbit of Pluto.

You might not know it, but our entire solar system is enshrouded by a giant cloud of large dirty snowballs hanging out around 5 to 10 trillion miles from the Sun. It's named the Oort Cloud, after the man who hypothesized its existence, and is home to millions of comets that are ever-so-loosely bound to the Sun. With such a tenuous attachment to the Sun, these objects can be disturbed by the motions of other, relatively nearby stars.

Yes, they are so far away from the orange in Houston that the orange in San Francisco can actually perturb them, and yet they are associated with the orange in Houston.

If they do get jostled out of their happy Oort Cloud existence, these things have an ugly tendency to come crashing into *our* neighborhood. We saw this kind of thing in 1996 when Comet Hyakutake appeared unannounced in our sky, zipped past Earth and around the Sun, and scooted right back to the outer solar system. Okay, it wasn't quite that fast, but it was certainly a brief visit for a comet.

Hyakutake was discovered through binoculars on January 30, 1996, by Yuji Hyakutake when it was only 200 million miles from the Sun. It's not just coincidence that the names are the same. Comets are the only astronomical objects that you can actually have named after you while you're still alive.

"That's not true," you might protest. "I just bought a star for my cousin Bob. I've got the paperwork right here."

I have bad news for you. You can't actually get a star named after your favorite cousin, even though several unscrupulous individuals will let you "buy" a star for the low, low price of $49.99 and even sell you the official, leather-bound "star regis-try" for only three payments of $34.99, thus proving that the star formerly known as HD 25329 is now called Bob. Don't fall for these scams. It's better to give your $49.99 and three payments of $34.99 to your local planetarium, which might actually give you the opportunity to "adopt" a star in Bob's honor. They'll give you official-looking paperwork, too, but they'll never try to convince you that it's anything more than a fundraiser.

Now if you're dead set on naming an astronomical object Bob, it is possible. Just be the first to announce the discovery of a comet. And be named Bob. Most likely you'll share the discovery with someone else, though, so it'll wind up being Comet Bob-Marley or something.

So anyway, as I was saying . . . Comet Hyakutake was discovered when it was a mere 100 million miles from Earth and closing fast. Within two months, the comet was less than 10 million miles from us, and if you had been a patient enough observer, you could have actually detected its motion against the distant stars. In fact, for a few days in late March 1996, it was one of the brightest objects in the sky, easily visible even in the glow of downtown city lights. The comet itself was pretty scrawny by comet standards—just a 2-mile-wide hunk of dry ice and dirt—but its gaseous tail was over 50 million miles long, stretching so far across our night sky that you needed a panoramic camera to capture all of it.

It was really, really cool.

And, frighteningly, it was really, really close. Now 10 million miles might seem comfortably distant to those of us who think a walk around the block is a long way, but in solar system terms, this is a hair's breadth away from us. If Hyakutake had been a dart aiming for us on the great solar system dartboard, it would have scored 50 points for a cosmic bull's-eye. And despite its small size, if it had actually hit us, it would likely have caused a global catastrophe and mass extinctions, all thanks to a minor disturbance to our Oort Cloud hundreds of thousands or even millions of years ago.

Fortunately, those sorts of disturbances aren't as frequent as they could be. Remember that the Sun is a considerate ice-skater, cruising along the Galactic ice rink in a fairly circular path like most of the other stars in our neighborhood. All this good behavior thankfully helps to keep most Oort Cloud objects safely in the Oort Cloud where they belong. It's sobering to realize that for every Hyakutake flung in our direction, there are countless others that happily stay out in the outer solar system simply because the Sun is sticking with the local Galactic traffic pattern.

Just as your driver's ed teacher always said, it's important to go with the flow.

But even keeping with the flow of traffic isn't enough. There's also, as it turns out, the whole cosmic timing issue (forget for a moment that time is just an illusion). Recall that chemistry students in the Early Universe—if they could have existed—would have had to memorize only hydrogen, helium, and lithium before they could ace the exam on the periodic table of the elements. These millennia, the Universe boasts over ninety naturally occurring elements, making chemistry exams a bit more challenging. And with the exception of the first three elements, all of them were manufactured inside stars. Even as you read this, stars are busily making more and more of those heavier elements and using up hydrogen, the lightest one.

If you look at the elemental inventory of Earth, you'll see that we've got huge stores of iron and nickel and oxygen and other things that at least two generations of stars have had a hand in making. Even a Sun-like star in the right place in the Galaxy at the wrong time in cosmic history couldn't have supported an Earth-like planet. And evidence seems to indicate that as stars form from clouds that have even higher concentrations of the heavier elements, their inner planets are more Jupiter-like and incapable of sustaining life, at least life as we know it. Billions of years from now, an earthy planet at the right distance from a Sun-like star might be a thing of the past.

Only a small percentage of stars that have called the Milky Way home has been (or will be) in the right place at the right time. And a small percentage of *those* are single stars with the right mass. And around those stars is a narrow region where conditions are suitable for creating Earth-like planets that you could comfortably call home. The chances of making an Earth get slimmer by the second, right?

Yes, but remember that our Galaxy has over 100 billion stars— suns—and there are literally billions and billions of galaxies out there in the Universe. The numbers are truly staggering. In fact, they're *astronomical*. Even when you factor in all the small percentages of small percentages of small percentages, this puts the number of potential life-harboring planets in the tens of thousands to millions.

Still, it looks pretty lonely from where we're sitting. So let's take a moment to appreciate home's little quirks.

6.4 Why Goldilocks Could Never Have Been a Successful Real Estate Agent

Once you've returned mosquito-bitten and sunburned from your camping trip, take a good long look around your home. All the exasperating flaws probably look pretty minor right now because you're *home*. The bathroom is nearby; the beds are soft and free of bugs; it doesn't usually rain on you while you're preparing dinner. It's practically paradise. Okay, admittedly last week you were threatening to tear the whole place down, but the problems with home suddenly seem friendlier. Somewhere, deep inside, you realize that it just wouldn't be home without them.

Now that you've gotten a taste of the perils away from our planetary home, you are probably saying the same for Earth. Sure, there are plenty of things you might want to fix about this place. For instance, earthquakes and volcanoes might seem to be pretty nasty features of an otherwise idyllic planet. I'm sure the residents of Pompeii would agree that they'd have preferred life on a planet that didn't produce toxic showers of flaming ash without warning. Ironically, the very thing that makes for deadly volcanoes like Vesuvius and tsunami-producing earthquakes also appears to have a hand in keeping this place livable.

Blame our weak foundation for those. You spend your life on a paper-thin layer of granite that we laughingly call *terra firma*. It's anything but. This tiny shell of rock, floating upon oceans of molten material, ranges from just a few miles thick at the deepest trenches in the ocean to a thickness of over 40 miles in the Himalayas. Compare this to the 3,000-plus-miles-deep cauldron of molten death beneath the crust and our solid rocky foundation is starting to look frighteningly shaky.

The shell itself isn't even solid or stable. It's broken like a jigsaw puzzle into ginormous pieces called plates that are driven around the surface by the churning lava lamp–like motions of the interior. The overall process is called plate tectonics. A more familiar term is continental drift, which sounds so light and fluffy, like a snowflake settling onto the head of a furry woodland critter. The reality is more violent and destructive, like a magnitude 8, tsunami-producing earthquake resulting from continent-sized plates of rock scraping against each other.

Continental full-contact gauntlet-of-destruction would be more like it, and you might be tempted to think that life on a planet without this chaos would be preferable. Some place like Mars, for instance. Lately, though, it's become apparent that without this chaos, life wouldn't be here at all. You see, plates don't just bump around on the surface ("Pardon me . . . excuse me . . . oops, did I cause that mountain range? I'm so sorry!"). Sometimes one plate is literally forced beneath another. The solid crust of the submerged plate then becomes part of the molten, circulating interior. Elsewhere, as other plates separate from each other, molten material seeps through the thin, weak spot and hardens to create new crust. In this way, old, "used" crust is replaced by new and improved crust, complete with a fresh set of nutrients for life on the surface.

Sure we've got a cracked foundation, but we've got a heck of a recycling program!

On top of this, the constant crustal turnover helps regulate what's in our atmosphere. Life forms like bacteria pull carbon from the air and deposit it in the solid crust. As plates jostle around, this carbon is later outgassed as carbon dioxide by volcanoes in a giant planetary belch. This keeps the atmospheric carbon dioxide levels pretty stable over millions of years and, because carbon dioxide is a greenhouse gas, this means our global temperature stays pretty stable over millions of years as well. In fact, it's been suggested that the lack of plate motion on Venus contributed to its hellish conditions.

Score another point for unstable foundations.

So what drives these plates around? Think lava lamps. Awesome retro lava lamps with the incandescent light bulb in the base and the horror movie–inspired red blobs of . . . what the heck *is* that stuff, anyway? Whatever it is, and whatever that clear liquid is that they're floating in, they do a great job of illustrating the goings on in the interior of our planet.

Now if you were a child of the 1960s or 1970s, you might be inclined to think that the light bulb was placed on the bottom for psychedelic reasons. Turns out that it won't work with the light bulb at the top (try turning one upside down). The light bulb heats the blobs and the clear mystery fluid. This hot material is less dense than the cooler stuff at the top of the lamp, so it floats to the top while the cooler stuff sinks to the bottom. Then the formerly cool material gets to the bottom, heats up, and rises, while the formerly hot material cools off and falls, and voilà! A convection cell is born.

The interior of Earth is just teeming with these sorts of things. Convection cells, not lava lamps. The innermost part of Earth is a solid, highly compressed ball of iron about the size of the Moon and insanely hot—far hotter than the hottest black asphalt on a scorching summer day. In fact, it's about the same temperature as the surface of the Sun. That heat energy's got to go somewhere, so it hitches a ride on the molten rock and metal, which rises and transfers some of the heat to the outer layers. Meanwhile, billions of years of radiating heat into the coldness of space has helped harden the top few miles of our planet. Earth's kind of like a giant sphere of fried mozzarella in that respect—crunchy and cool on the outside, but hot and gooey on the inside.

With all the rising and falling convection cells, the light, fluffy plates on the crust have no choice but to be shoved around. In one place they might slam head on, creating a mountain range like the Himalayas, which is growing taller by the day. Meanwhile, in the depths of the Atlantic, plates are gradually separating, causing the sea floor to expand and making the trip from South America to Africa an inch or so longer every year. In places like California, plates scrape alongside each other and produce epic earthquakes.

'No problem,' you might think. 'I don't live along any fault lines.' You might be fine for the duration of a human lifetime, even on a fault line. Eventually, though, everything on the surface is at the mercy of Earth's churning interior. Depressing as it might seem, the pyramids themselves will eventually be melted down.

On the other hand, so will all evidence of Paris Hilton, so it's not all bad.

In fact, the bad is far outweighed by the good, even before you consider the fate of embarrassing pop culture icons. That churning interior, along with Earth's relatively quick rotation rate, is most likely the engine behind our magnetic field, which in turn helps keep you alive long enough to perform the interpretive dance to learn how it manages such a feat. Actually it's more like performance art than interpretive dance, and it requires a few simple props: a small length of wire, a 9-volt battery, and a compass. No, not the pointy thing that frustrated you endlessly in math class as you attempted to draw a circle, but the one that helps you find north. Once you have these things in hand, it's child's play to create your own magnetic field that dwarfs Earth's, at least on a very localized scale. All you have to do is hook one end of the wire to one terminal of the battery. Then hold the compass above the wire while you briefly—*briefly*—touch the other end of the wire to the other end of the battery.

Briefly.

If you keep it there too long, your battery and wire will get uncomfortably hot. Searing hot. Set-things-on-fire kind of hot.

So don't do that.

While you're in the act of closing that rather uninteresting circuit, though, take a look at the compass needle. When you touch the other battery terminal and allow electricity to flow through your wire, the needle will find a whole new "north." As you just found out, and as was discovered accidentally by a very surprised Hans Christian Ørsted in 1819, flowing electricity creates a magnetic field. With a few household objects, you've discovered the principle behind electromagnets, your car's ignition system, and a global magnetic field that shields all life on this planet. Not bad for a 30-second performance art segment.

Obviously there's not some giant 9-volt battery connecting Earth's poles, but there is definitely a way for it to get electric current flowing. First it needs something metallic so that the charges can be pushed around like so many 98-pound weaklings at the beach. The iron-nickel core does nicely. Then something actually needs to be pushing those charges around. This is where the fluid convection cells and fast rotation rate come in handy.

Earth's molten, metallic interior and rotation rate apparently

A fragile, self-renewing home

New crust

Earth's rotation

How thin IS Earth's crust? This line is roughly to scale, $\frac{1}{200}$ the distance to the center.

Flow in the puttylike mantle pushes the crust around in thin sheets

Lava lamp (not to scale)

Mars-sized liquid outer core has flow pattern driven by heat, rotation, and influence of the Moon.

Moon-sized solid iron-nickel core rotates one to two degrees faster per year than the planet's surface. Together these appear to generate Earth's magnetic field.

keep everything churning in just the right way to produce and maintain a relatively stable magnetic field. Sure it has its polarity reversals, pole migrations, and periods of strengthening and weakening, but Earth's wa-ay better off than Venus or Mars. In fact, this probably explains why Venus never could produce its own global magnetic field. Rotating at such a leisurely pace, all its molten metal simply doesn't get wound up enough to make a planetary electromagnet. As for Mars, its tiny size meant that it cooled off faster. This allowed the crust to get so thick that there's nothing churning around inside it anymore, despite the fact that it's rotating at the same rate Earth is. In its past, Mars probably produced a magnetic field in much the same way Earth does, but right now all it has are small leftover patches of magnetic rock here and there.

Once again, there's no place like home.

But it's still fun to go camping every now and then to get a finer appreciation for it. S'mores on Mars, anyone?

SMALL WONDER

Looking for a Home Away from Home

The relative rarity of Earth-like worlds hasn't stopped astronomers from attempting to locate them. Since 1995, we've been detecting planets around other stars, a pretty amazing feat considering how tiny planets are and how huge and luminous stars are. Unfortunately, for the most part, we've only been good at detecting hugely massive planets in tiny orbits around their stars. These hot Jupiters are the easiest things to find because they have the most obvious effects on their stars on the shortest time scales, but they're nothing like home.

All that might change. In March 2009, NASA launched the

Kepler mission, whose sole purpose in life is to check for the existence of Earth-sized planets in the so-called habitable (i.e. "just right") zones of stars (see http://kepler.nasa.gov/ for oodles of information on this mission). The idea is basically the same as an eclipse or an occultation, but much harder to spot. From our perspective, a planet in another solar system will sometimes pass in front of its parent star. When it does, the star will appear to dim slightly.

Very slightly.

For a whopping big planet like Jupiter passing in front of the Sun, an alien observer might expect to see the Sun's brightness dim by about 1% for 30 hours every 12 years. Earth would block less than 1% of 1% of the Sun's light, and the dimming would last all of 13 hours once a year.

Imagine watching the Sun for an entire year to see a minuscule dimming that lasts just half a day! And it only works if you're looking at our solar system from the correct angle where Earth appears to pass in front of the Sun. The chances of being oriented at the right angle to witness that are less than 1%.

Yet this is exactly what the Kepler folks are looking for. Again, even though there are fractions of fractions of a percent involved, astronomers are hoping to capitalize on the sheer number of stars in our neighborhood. By monitoring 100,000 stars, they figure they should be able to detect anywhere from 50 to 200 Earth-like planets in their stars' habitable zones. This should be doable within its four-year mission, considering that within the first nine days of operations, Kepler discovered five Jupiter-like planets.

And then what? Even more ambitious missions are in line to study the very atmospheres of these planets for telltale signs of life. Still other projects are looking for evidence that intelligent life is out there.

 **SMALL WONDER**

Do Aliens Exist?

Maybe.

 **SMALL WONDER**

No, Really. Do Aliens Exist?

Think about it. There are jillions of planets out there, and the basic ingredients have been coughed out by stars all over the Universe. It wouldn't be crazy to think that some of those stars have planets that are home to *something*. And maybe that something is reading about how wondrous its own home is.

If, on the other hand, the question is whether aliens with a slightly modified version of the human body—you know, two hands, two feet, oversized human-ish skull—are hanging around Earth and periodically mutilating our cattle, I'd have to say "no." For one, it seems a huge waste of their time. Seriously. They traipsed across vast interstellar distances to get here and do what? Play hide-and-seek with us and do bovine vivisections? Would *you* take a generations-long trip to a completely unknown place just to confuse the locals? For another thing, just look at the bizarre variety of creatures here on our own planet. That in itself should convince you that something arising on an entirely different planet should be as different-looking from us as we are from those creepy looking fish that hang out at the bottom of the ocean. Just take a look at the Census of Marine Life at http://coml.org/imagegallery to find out what otherworldly things are sharing our planet with us. My guess is that the real ETs out there in the Universe are unlike anything we can imagine.

It kind of makes you wonder what *they'd* think of *us*. And Paris Hilton.

SMALL WONDER

Wanted—Alien Hunters

So you're still determined to find some aliens? Fortunately, the good folks searching for extraterrestrial intelligence (a.k.a. SETI) could use your help. They've been scanning the skies with radio telescopes since 1960 and have amassed quite a pile of data. Now all they need to do is sort out the "natural" signals from the suspected "non-natural" signals.

That might sound impossible, but radio astronomers have a pretty decent idea about what constitutes a natural radio wave signal from astronomical objects and what would likely be something produced artificially. We even sent our own radio signal into space in 1974. It will arrive at its destination, the star cluster M13, around the year 26,974, give or take, at which time the computers at the radio observatory on some alien planet will chime, "You've got mail!"

Astronomers figure that if we were hopeful enough to send an electronic greeting card to the Universe, then there's a decent chance that someone out there might have been inclined to send out its own message. The problem is that we've got to dig through mountains of other stuff before we find the one nugget of gold, the radio signal that screams out "Intelligent Lifeforms R Us." That's where you come in. It takes colossal computing power to analyze all this data. More than all the observatories combined. So since 1995, the data have been farmed out to people like you and me who have home computers that sit idle most of the time. The program's called SETI@home (http://setiathome.ssl.berkeley.edu/), and it uses millions of sleeping home computers to sift through the data in the search for artificial signals.

So far nobody's struck gold, but the digging has barely begun. Good luck.

CHAPTER 7

Wonder

Wisdom begins in wonder.
—Socrates

S o here you sit, the product of 13.7 billion years of cosmic evolution, reading about the last wonder of the Universe. Unlike the other wonders, this one actually resides within you. Even more impressive is that it is quite likely the most inexplicable and wondrous wonder of them all: wonder itself.

Have you ever wondered why you wonder? I know . . . you've got More Important Things to do. If you're reading this, you were most likely told so many times to get your head out of the clouds and be more responsible that you've forgotten that iridescent bugs are cool looking and that you should stop what you're doing and stare at the Moon whenever you see it and that you should ask "why" about everything you encounter.

In short, you've forgotten how to look at the world like a preschooler, and that's a real shame because everyone knows that cleaning the kitchen and taking out the trash aren't particularly edifying. Nobody needs to tell you there's nothing mysterious or wondrous about paying bills. Even in the midst of these rather dreary tasks, though, you should force yourself to look at them as only a small child can: with a sense of wonder. After all, even a monkey can be trained to take out the trash, but only you are uniquely qualified to appreciate the chore on a level the monkey never will.

But why?

7.1 Life—The Ultimate Eating Machine

Let's recap a bit before tackling a question as profound as, "Why can't a monkey experience the awe of taking out the trash?"

First off, the wonder of *light* was generously converted to matter after the Big Bang (the leftover light of which can still be observed in the long wavelengths of microwaves), and because of a somewhat mysterious preference for matter over antimatter, the Universe is now filled with the wonder of *stuff.* This is pretty convenient because you, too, are made of stuff. Some of your stuff—specifically your hydrogen—is even still in its pristine state, unperturbed since the Universe was just minutes old, although its electron has probably been restocked countless times.

Fast forward through the next 9 billion years (a feat you can accomplish only through the wonder of *time*) and the Sun's ancestors have diligently manufactured and distributed into space the rest of the elements that everyone painstakingly ignores on the periodic charts in science class. These elements made up the ginormous cloud from which the Sun and solar system—and likely dozens of other stars and solar systems—formed nearly 5 billion years ago.

Another wonder—*gravity* (probably with the aid of a shock wave from a nearby exploding star)—took over and pulled the cloud into a massive, hot, central ball with an entourage of planets, which also owe their existence to the wonder of gravity. Most of these planets were either too hot or too cold, too massive or too scrawny, but the conditions on our wondrous *home* planet, with its alternating cycle of day and *night* (another wonder), were just

right for something called life to get going and keep going until it began asking questions about itself.

But what is life? Like time, it's one of those things that keep scientists and philosophers awake at night. Renowned physicist Erwin Schrödinger wrote a book by that very name—*What Is Life?*—in 1944. Interestingly, this is the same guy who posed a hypothetical feline life-and-death scenario to describe the mysterious nature of quantum mechanics, the rules governing behavior on the smallest of scales. Since you've probably heard of Schrödinger's cat, but were always afraid to ask what it was, here's the basic story. Schrödinger asks us to suppose that a cat is placed in a box with a vial of poison, a Geiger counter, and a tiny bit of something radioactive like uranium, an atom of which might spit out a bit of radiation. Or it might not. It's hard to tell. Atoms are like that, and you just can't predict what they'll do next. This is in complete contrast to the situation with a falling ball, where you can describe its behavior, both past and future, with amazing precision. That's the whole point of quantum mechanics. It's a kind of guessing game with lots of statistical chances of events, but no certainty.

Anyway, if an atom of the uranium decays and releases radiation, the Geiger counter clicks and triggers something to break the vial and the cat dies. (Don't run off and call the ASPCA. It's a hypothetical situation. I'm betting, though, that Schrödinger wasn't much of a cat person.) If the sample doesn't decay, the Geiger counter stays silent and the cat lives. Quantum mechanically—and completely nonintuitively—if someone asks you whether the cat is dead or alive, you have to declare the cat simultaneously half-dead and half-alive. Moreover, if you peek, you ruin the whole thing and, in the lingo of quantum mechanics, the act of peeking causes the cat's wave function to collapse into exactly a dead state or an alive state.

Call me crazy, but it seems a bit ironic for a guy who can glibly discuss a state of simultaneous life-death to write an entire book about what life is. But it was a rather insightful book, and it was loaded with concepts that the biology folks would only later confirm. What's really amazing is that it took him an entire book just to attempt to answer a simple question. And scores of books were written on the subject before Schrödinger tackled it, and scores more have been written since.

Clearly life is more than just a pile of ingredients. Mixed together in a giant bowl, 95 pounds of oxygen, 35 pounds of carbon, 15 pounds of hydrogen, 4 pounds of nitrogen, 2 pounds each of calcium and phosphorus, a cup of this and a pinch of that would probably not look or act anything like you. Somehow *you* have a property that organizes those 150 pounds of goop in such a way that you can eat, breathe, move, and play with an iPhone. There's something about *you* that makes you fundamentally different from a cauldron of chemicals. In the immortal words of the late Carl Sagan, "The beauty of a living thing is not the atoms that go into it, but the way those atoms are put together." (For a great synthesized re-mix of Carl Sagan saying these very words amid some rather . . . um . . . interesting music, go to www.symphonyof science.com and scroll down to "We Are All Connected.")

Given the entropy discussion in chapter 5, you might think that being an organized life form goes against everything the Universe is about. After all, isn't everything supposed to go from an ordered state to a disordered state?

Once again, I've lied to you. Well, maybe not lied, but I've possibly misled you. You see, it's not so much about the physical arrangement of things—a broken plate versus an unbroken plate, for instance. It's about Universal laziness. The Universe is an inherently lazy place, and what it really prefers to do is whatever takes the least effort.

At nearly 14 billion years old, it is apparently the largest and most ancient cat.

The great thing is that you can get a hint of this Universal laziness with something you probably already own. All you need are some refrigerator magnets.

Go ahead and get them. It's the closest thing to an interpretive dance you'll get to do for this chapter.

Got them? Good. Now try to stick them together.

"Um," you say with furrowed brow. "This explains life?"

Well, not completely, but it's a good start. You see, although they're probably not labeled with their north and south magnetic poles, you can figure out pretty quickly when you're trying to force two of the same poles together. They resist your attempts to make them touch. If you were as lazy as the Universe, you wouldn't bother putting in the effort of forcing these things together. But when you place the opposite poles near each other, it takes effort to keep the magnets *apart*. So the Universe just lets the magnets join together, which is what they wanted to do in the first place.

I can see the comic books now: "Captain Universe! When he's near, magnets stick together—all on their *own*!" This new-and-improved Captain Universe would be a pretty huge disappointment, lazily sitting back and watching everything around him fall into its natural place. Come to think of it, that sounds strikingly similar to the ancient Greeks' view of the world. Maybe they weren't so far off after all, but once again, we "modern" scientists have found a way to package the concept in more sophisticated terminology.

This Universal laziness clause also applies to chemistry. Let hydrogen and oxygen atoms mingle with each other, add a bit of so-called activation energy (a hot match, for instance), and they'll quite happily come together to make water molecules. In the process they'll also release some energy. (Note: This is *not* fusion. I repeat. This is *not* fusion. It's just a chemical reaction, which, when compared to fusion, is a bit like comparing a small penlight to one of those enormous search lights at your nearby car dealership.) Just as in the case of the magnets, it takes less effort to be a water molecule than to be independent hydrogen and oxygen atoms, so they naturally combine and give all that leftover energy back to the Universe.

At first glance, it might look like you've got a more organized situation, but the reality is that, energetically speaking, things are worse off than they were before. All the extra energy is now useless for the most part. In scientific lingo, water is lower in thermodynamic energy than its separate atoms, hydrogen and oxygen, and the Universe takes advantage of the chance (in this case, some outside activation energy) to get to that lower energy place.

It's just that lazy.

Water isn't the only thing that can be made rather readily from its component atoms. Lots of molecules come together spontaneously, even in the giant clouds of interstellar space. The most abundant organic molecules out there are called poly- cyclic aromatic hydrocarbons, which sound like there should be a fundraiser to combat them. The more familiar term, at least to the astronomers who study them, is PAHs, and these things have oodles of carbon and hydrogen and bits of other things all linked together in complex molecules that resemble, of all things, soot. Universal laziness gets these atoms congregated into something that, with a tiny bit of chemical tweaking, can actually help run the photosynthetic energy factories in plants.

Giant molecular clouds also host sugars ("What a sweet Uni- verse we live in!" you shout), along with such delightful-smelling chemicals as ammonia, methane, and formaldehyde ("Okay, maybe not."). But what is probably most surprising is that astron- omers have found actual amino acids floating around in space.

"No kidding!" you exclaim.

"What's a mean ol' acid?" the child asks.

Amino acids are actually anything but mean. Without them, you wouldn't be doing anything right now. They're the building blocks of proteins, which are complex molecules that pretty much run life as we know it. In fact, your genetic code, which programs everything your body does, is nothing more than huge strings of amino acids all arranged in a particular order. You personally have twenty varieties of amino acids in you, a feature shared by all life on Earth, but lots more exist. Monosodium glutamate, for one, which makes the Universe a more flavorful place to live.

Although your kung pao chicken probably has plenty of MSG, you don't owe your life to it. An amino acid that you *do* owe your life to, though, is glycine. In 2009, this amino acid was detected on Comet Wild 2 (that's not pronounced "wild," as in "wild and crazy," but pronounced "vilt," as in . . . um . . . "vilt"). Considering the fact that comets have been striking Earth for several billion years, it's not so vilt and crazy to imagine that many of life's building blocks were supplied by comets, which lazily scooped such raw ingredients up from the solar-system-forming cloud and dropped them off here like so many interstellar storks leaving new babies.

Not all of life's scaffolding is the result of Universal laziness, though. Some things actually require outside help to exist. Plants, for instance, continually make use of the energy in sunlight as they organize their photosynthesizing lives. We, on the other hand, have to take advantage of the organizational abilities of plants—and other creatures that take advantage of plants—to keep all our molecules in good working order. While you personally aren't heading to a state of minimum thermodynamic energy (a really cool-sounding term for death), you're certainly helping the Universe along in that respect. You take in organized molecules (a not-nearly-cool-enough-sounding term for chocolate-covered strawberries), shred the poor little things, rearrange and assimilate their atoms so that your molecules keep doing what they're doing, and then get rid of the leftovers. Along the way you give off lots of heat, which is nothing more than your body's way of dumping all the useless energy into the rest of the Universe and temporarily keeping itself from heading down that same energetically disordered path.

In a strictly physical sense, "life" is little more than the amazing ability of molecules to tap into the energy of their surroundings and kick off waste energy and molecules. In other words, we're eating machines. But we're not the only things that use fuel. The flame on a candle sustains itself by combustion, shedding loads of heat in the process. Nobody would call it alive, though. Something more is happening in this process we call life.

7.2 There's More to Life Than Eating

That statement would come as a great shock to a teenage boy, but it's true. If you base your definition of life strictly on its ability to siphon energy from its surroundings, then you've got to include candles and cars. Another feature that sets life apart is that living things are amazingly good—but not perfect—at making copies of themselves, something that neither candles nor cars can do.

Once again, life isn't the only thing that's good at replicating itself. You can find things perfectly willing to replicate themselves in your own kitchen. No, I'm not talking about the ecosystem in the back of your refrigerator. I'm talking about simple crystals. In an amazingly easy and tasty process, you can grow your own rock candy. You can find the recipe on the Internet and test the patience of a child for the week or so that it takes to grow a decent hunk of highly organized sugar molecules. Rock candy is just another example of Universal laziness.

Even though they can grow and reproduce using available resources, sugar crystals aren't remotely alive. Still, they might hold some clues about the origins of life. You see, the sugar crystals don't require any special circumstances to grow. No special one-time event is needed to create rock candy. It's easily done in any kitchen with the right combination of ingredients. Crystal growth is, under the right conditions, inevitable. It's even possible that life itself is, as well.

This isn't about the big stuff yet. Sure, back in the 1600s people honestly thought that you could create fully formed male and female mice by leaving some grains on a piece of dirty, moist clothing for three weeks. Or, if you were in the market to create a much more disgusting life form, you could just leave meat out in the open. Voilà! Rotten meat with a nice garnish of maggots.

Yummy with a side of yummy sauce!

In a simple, yet powerful, controlled experiment, a fellow named Francesco Redi—who must have had an incredibly strong stomach—decided to check on the premise of these recipes in 1668. He put some meat out in the open and let it rot. Sure 'nuff, maggots appeared.

Of course, everyone *knew* that's what would happen because, as far as anyone could tell, that's how maggots—and ultimately flies—came into existence. But then Redi tried something really novel. He covered the meat, either with a mesh screen or airtight seal. To the surprise of most people, the rotting meat that was off-limits to egg-laying flies never created maggots, and the idea that life could spontaneously erupt from a nonliving thing rotted away like so much unrefrigerated meat.

But this isn't the sort of life-from-nonlife that biologists discuss these days. To get an idea about life's true origins, you have to think smaller than maggots.

Far, far smaller.

Smaller still.

As it turns out, we humans are pretty chauvinistic about life. Ask a random person to name any life form, and you'll probably hear about birds and trees and wallabies. Maybe maggots. Almost nobody thinks of the tiny bacterium *Prochlorococcus Marinus* (here "tiny" means that its name on the page is about 100,000 times longer than it is. Unless your name is spectacularly long, your name-length-to-height ratio is probably considerably smaller). It turns out that *Prochlorococcus Marinus* is responsible for about half of the photosynthesis in the ocean and about a fifth of the total photosynthesis on planet Earth. Yes, photosynthesis, the process that turns sunlight and carbon dioxide into usable fuel for the organism and kicks oxygen out as trash (again, the Universe is all about taking out the trash). We're actively inhaling the waste product of this unassuming bacterium all the time, and I'll bet you've never once paused between your 20,000 daily breaths to thank it. Instead, we thank our philodendrons and sequoias because, hey, we can *see* those.

Anyway, life on smaller scales is pretty different from your daily existence. The *Prochlorococcus Marinus* and other microbes spend their days taking in nutrients, spitting out the stuff they don't need, and making copies of themselves, sometimes at incredible rates. Within a single day, given the right growing conditions, the "offspring" of a single bacterium can number in the billions of trillions. (Note that this is not billions *or* trillions but billions *of* trillions! Fortunately, this is mostly a mathematical exercise to demonstrate geometric growth, so the reality isn't quite that enormous.) Although they're slaves to a rather boring routine, they are still *alive.*

Frighteningly, even these "simple" things are insanely complicated compared to the sugar crystals that you can grow in your kitchen. What we consider "life" shows up somewhere between the 24-atom molecule that is table sugar and the hundred-trillion-atom cooperative that makes up the cell structure of even the simplest bacterium.

Fortunately for biological science, there are some midsized complex molecules that can make copies of themselves as they do the Universal lazy dance. One tantalizing example is something called a ribozyme. The underlying structure of a ribozyme is the substance ribose, which is a type of sugar molecule. All sugar molecules have names ending in "ose," lots of which are probably familiar to you. There's glucose, sucrose, fructose, lac-

tose, dextrose, decompose, overexpose, and pantyhose. (Note
to self: Check on those last few to confirm they are sugars.) The
"zyme" part tells us it's an enzyme, which is just a protein that
helps certain chemical processes run faster.

It sounds like a pretty basic molecule when you put it like that,
but a quick jaunt to the Exploring Life's Origins website (http://
exploringorigins.org/ribozymes.html) will show you that it is, in
fact, horrendously impossible to draw, despite the illustrator's
assertions to the contrary. Although it appears pretty complex,
a certain type of ribozyme placed in a solution filled with the
appropriate raw materials will replicate the heck out of itself.
Then you'll have a whole swarm of things that you can't draw.
Even more impressive is that, if you leave the solution in sunlight
or otherwise expose it to some outside energy source, you'll wind
up with some slight variations in your original pattern.

In other words, these molecules evolve.

I know. It's as politically dangerous a term as the greenhouse
effect, but this, too, has a very straightforward meaning. Evolution is nothing more than a process of changing over time. Car
designs evolve (with some help from the designers, of course).
Your musical preferences evolve. And there is more than enough
evidence that life itself has evolved and is still evolving. Heck,
bacteria and viruses evolve so quickly that we have a bear of a
time trying to kill them. We think we've found a great antibiotic
and the next thing we know, those pesky bacteria have developed
a resistance to it.

So what does a happy ribozyme evolve into? Apparently a
slightly different ribozyme. The self-replicating process isn't perfect, and occasionally an atom will go where it wasn't originally
programmed to go. The Universe doesn't always put the round
peg into the round hole, a problem that arises because the pegs
and holes are zipping around at hundreds or even thousands
of miles an hour. Occasionally mistakes—or, rather, variations—
happen as an oval peg is jammed into that round hole. What's

really cool is that the new ribozyme design might or might not be better at copying itself than the old ribozyme design. If it *is* better, then it gobbles up the raw materials faster and keeps the old design from making as many copies of itself. Before you know it, a whole new ribozyme is king of the pond.

But ribozymes aren't technically alive, either. Somewhere, through pathways that are not fully understood, complex self-replicating molecules found themselves within a protective barrier. This setup was pretty handy for keeping other self-replicating molecules from ripping apart *your* self-replicating molecules and playing keep-away with the fragments ("Mo-o-om! Jimmy's pulling my proteins off again!"). If you wanted to keep up with the Joneses at this point in life's history, you needed to have a protective barrier, too. Then you had to replicate not only your huge molecular chain, but also your protective bubble, which we call the cell wall. Once you could reproduce an entire cell—wall, innards, and all—you became officially "alive" from the somewhat arrogant standpoint of some multicellular critters billions of years later.

Being a one-celled organism is definitely popular. For most of Earth's history that's pretty much all there was. Even now, single-celled life still dominates, a fact that is easy to overlook since the individual beings are so vanishingly tiny. Pull out a cheap microscope, though, and you'll see all kinds of fun things living in pond water. Your own body is even home to zillions of bacteria that, despite the ick factor, typically make your life better.

Being a single-celled thing clearly works, evolutionarily speaking, or microbes would have all been replaced by something new and improved somewhere during the nearly 4 billion years that life has been on our planet. But not all single-celled creatures are created equal. It was only a matter of time before something popped into existence that had the right molecular tools to rip apart those first single-celled things and make use of their ingredients. Then something *else* came along and one-upped that thing. For a good 2.5 to 3 billion years of the 4-billion-year history of life, single-celled organisms developed a stunning array of skills to exploit the available resources (often other single-celled critters) to stave off minimum thermodynamic energy.

And then something truly novel occurred. About a billion years ago, single-celled organisms began banding together. This "circling the wagons" approach of early cell colonies made the chances of aggregate survival even better. Collections of cells with different talents ("I'll pump the water if you take out the trash") ultimately formed, and by half a billion years ago, Earth witnessed the emergence of the first relatively complex forms of life. Then, in a prehistoric version of Xtreme sports, some of these things even held their breath and briefly emerged from the waters, leaving their slug trails (or possibly prelobsterlike prints—it's hard to tell) in the mud for modern paleontologists to find.

As it turned out, life up top was a mixed bag. There were no predators (good), but, tragically, no food either (bad), because plants hadn't yet made the trip, unless you want to count some primitive mosslike stuff. A newly adapted land critter couldn't stray far from those dangerous watering holes if it wanted a snack. Still, it was a nice change of venue.

Pause for a moment here and imagine how our planet must have appeared to one of these intrepid explorers (assuming they could stop and take a look around). There would have been goopy, algae-ridden water wherever water gathered, the occasional giant sluglike or prelobsterlike critter wandering about, but not so much as a blade of grass on the land. There would have been freshly upthrust mountains devoid of pine forests, spring times without a single bird or flower or butterfly. For much of Earth's rich natural history, the picture has been nothing like what we envision when we yearn to get back to nature. On the other hand, mosquitoes weren't around, either, so it does have a certain appeal.

Fortunately for virtually everything we *do* associate with nature, green algae followed the intrepid "slugster" onto land fairly quickly (here "quickly" means about 100 million years, give or take a factor of two), bringing plant life to the surface. With fast food in the picture, it was only a matter of time before whole ecosystems sprouted up, including insects, amphibians, and reptiles inhabiting the newly developed forests and jungles.

Then, about 250 million years ago, nature's version of a crashing hard drive (or, possibly, massive glaciation) wiped out 95% of life on Earth, and there was no backup drive in sight. What life remained suddenly had lots of elbow room, a pretty fair amount of complexity, and very little competition. In a peculiar twist of fate, massive extinctions actually seem to have opened the door for rapid, extreme diversification and advancement, a feature not shared with a hard drive crash. In a mere 200 million years, mammals, birds, flowers, and many of the things we think of when we think of nature entered the scene and managed to survive the *next* massive hard-drive crash—the one that took out the dinosaurs and 50% of the species on the planet. With the giant predators out of the picture, tiny mammals started their takeover.

If you take a look around these days, you are actually witnessing a crazy and unprecedented variety of single- and multi-celled things that can swim, climb, reach, fly, slither, flutter, strangle, poison, and crush their way to the opportunity to make copies of themselves from the available resources, all while busily increasing the entropy of the Universe, which is all that those original molecules were doing in the first place. Apparently the goal of life is the same; only the packaging has changed.

It's probably a pretty safe bet, though, that none of those original molecules were *wondering* about any of this. Nor were the dinosaurs when they saw the giant fireball in the sky. Nor are the bacteria inhabiting your colon. Nor are the petunias in your flowerbed. In fact, for over 99.99% of the history of life on Earth, life has been pretty clueless about itself and about other things.

Until now, that is. As far as we can tell, we're the only beings ever to inhabit this planet who have had the capacity to wonder. The only ones! It's amazing, isn't it? What a privilege it is to be one of the elite creatures who can ponder the awesome Universal obligation to take out the trash, from the Big Bang onward.

And this means that somewhere along the way, the Universal laziness machine became conscious.

7.3 It's All in Your Head . . . Maybe

If you're like me, you're probably a bit insulted by the idea that you're nothing but a collection of atoms whose only "choice" is to behave the way they do based on the laws of physics. If that's the case, does that mean you and I have no more decision-making power than a ball caught in Earth's gravitational pit? Was I forced by laws of nature to write this book? And were you forced by laws of nature to read it? Was everything that you experience simply inevitable the instant that the Universe was born?

These are tough questions. Tougher than gravity. Tougher than time. Tougher than life itself.

So let's start with a slightly less tough question. Where do you live?

"Boston?" you reply with great hesitation.

Want to try again?

"Massachusetts? The United States of America? Earth? The Milky Way?!? Just tell me what you mean already!"

I don't mean any of those. What I mean is, Where does your "self" reside?

Most people, when pressed on this question, will say their head or their brain. And that's an interesting answer, because if there's anything you can't actually sense, it's your own brain. You can sense external light waves in the visible spectrum (in other words, you can see). You can sense compression waves in the air or another medium (you can hear). You can sense when your matter is interacting with other matter (you can feel, taste, and smell).

Perhaps because our eyes provide the window to the rest of the Universe we decide we must be residing *behind* them somehow, like a driver looking out the windshield of a car, and the brain just happens to be back there. But this concept of "self" wasn't always the case. Ancient Greek philosopher-scientist-politician-teacher-poet-ethicist Aristotle, who had quite a bit to say about pretty much everything, thought that the heart was the seat of your identity. As far as he knew, the brain's purpose was simply to cool the blood. While that might be true for some people you've met, it's a pretty weird concept for us modern sophisticates. Interestingly, when we discuss someone's innermost being, we still talk about what's in someone's heart, when we know that it contains ventricles and atria and not thoughts and emotions.

On the other hand, the brain itself seems to be nothing more than three or so pounds of cells interacting chemically and electrically with each other and with the rest of the body. Yet somehow we have this perception of self that *seems* to be more than just the brain, and it takes an amazingly short time for us to develop it. Even though other animals sometimes show uncannily human characteristics, it's unclear whether they are *aware* that they're aware. We *know* that humans are. This awareness of our awareness is called consciousness, because otherwise it sounds like we need to set aside some day in January to celebrate it. ("It's Awareness Awareness Day once again!") While we might celebrate it in a single day, understanding consciousness is a task that has taken centuries, and the answer still eludes us.

In case you've never taken a philosophy class, here's a brief snapshot of several millennia of philosophical studies of consciousness.

Witness...

Consciousness, as reflected in ancient Sumerian figurines, is a hunger to bear witness. We long to know with all our senses, and then to make new senses beyond the powers of flesh.

The ancient Sumerians (you remember . . . those 60 guys)
had an interesting cosmological take on consciousness, likening
it to a hunger to bear witness of the Universe *to* the Universe. In
other words, they seemed to understand that we were a product
of the Universe. They went a step further, though, and suggested
that we were, in fact, the Universe explaining itself to, well, itself.
We may not feel cut out for this task, but we're better suited to it
than anything else we know of. Galaxies, no matter how huge, are
incapable of comprehending anything. We, on the other hand, can
comprehend *them*. To an ancient Sumerian, human consciousness was the Universe's crowning achievement, the result of its
attempt to understand itself.

Good so far? Great. Let's skip a few millennia to Greek philosopher Plato, who came along and suggested that consciousness
pointed to the existence of another un-physical realm. To him,
the so-called real world was simply shadows of a higher perfect,
but not physical, reality. What we physically sense are imperfect
imitations of these higher forms, and the fact that we can do
something like *imagine* a perfect circle when we've never actually encountered one was evidence to Plato that we're connected
to this perfect reality. And, he argued, if we can tap into that un-
physical realm, then our "selves" have to be un-physical, as well.
On the other hand, the mind clearly had to have some interaction
with body parts or it could never actually make our bodies *do*
anything.

Moving along to Plato's star pupil, Aristotle, we find a guy who
wasn't too keen on his teacher's idea that there were perfect real-
ities in a spiritual realm. In fact, he was quite happy exploring
this imperfect natural world, so he suggested that the true reality
lay *within* the physical world, giving everything its nature. A sort
of infrastructure, if you will, that the physical world imperfectly
built upon. In his philosophy, humans (along with every other living thing) have a spiritual, intangible soul. It was something that
directs the movements of the stuff, but it was also somehow inter-

twined with it. For him, it was all about interpretive dance, the choreography and communication of countless molecules that becomes the essence of our "selves." And the director lived within the heart.

These sorts of mind-stretching arguments are precisely why I didn't major in philosophy.

You might have noticed that even Aristotle is no closer to providing a good explanation of consciousness. Since Aristotle, famous and not-so-famous philosophers have tended to fall under one of two main headings: mind-body dualist (meaning the "self" and body are separate things, although with obvious interactions) or physicalist (meaning that our sense of consciousness arises from something purely physical). These folks have brought up some tough questions, like "Can human consciousness survive bodily death?" and "What is the nature of consciousness?" and "Is free will a sensation or a reality?" and the ever-popular, "What is it like to be a bat?"

No, really, someone actually asked that one. It was Thomas Nagel, and the point was not to win the Random Question of the Year Award for 1974 (although I'm sure he did), but to demonstrate that no matter how much we know about the inner workings of the bat brain and how the bat brain interprets outside stimuli, we can never actually know what it's like to *be* a bat. Another analogy is that you can program everything about the color "red" into a computer and yet you'd never really expect the computer to *experience* red the way you do or I do or a guy named Steve in Sheboygan does. The sum of your lifetime of encounters with red, whether it be in stop signs or roses, cardinals or blood, will shape the way you respond to red in the future. Personal experience—and the "self" that arises from it—is not programmable.

Or perhaps it is. Recently studies in cognition and neuroscience have shown that there are situations where computers do mimic our perception, all the way down to our mistakes. It turns out that artificial neural networks have been "fooled" by the same optical illusions that people are fooled by, a result that lends support to the idea that what we perceive as reality is nothing more than a highly organized system of statistical likelihoods. 'The past 48,552 times I've detected this pattern, it has been a solid cube; therefore, it must be a solid cube now.' And then the brain is in for quite a surprise when it turns out to be something else.

It's not known whether the artificial neural networks then cry out (virtually, that is), "Wow! That's cool!" and then email the optical illusions to other artificial neural networks. They definitely don't spend money on M.C. Escher art and stare at it for hours while trying to puzzle it out.

Still, the brain appears to be nothing more than a giant computer with information received and organized in specific ways, ways that we can even mimic with human-made computers. Red light, for instance, will activate certain receptors in the back of the eye and be translated into an electrical signal that is sent to the occipital lobe in the back of the brain, which will register the signal as "red." If you damage the occipital lobe, you can experience visual hallucinations (seeing things that aren't really there) or loss of vision (um . . . not seeing things that *are* really there). All the while, memories and sensations are categorized and filed in the appropriate mental folders, tasks which continually modify our perception of the outside Universe.

And, just as a miscoded line in a computer program makes for some amusing "outtakes" in CGI movies (computer-generated imagery—think of productions by Pixar and Dreamworks), tinkering with parts of the brain sometimes produces truly bizarre consequences that aren't so much amusing as they are disturbing. Just read Oliver Sacks's *The Man Who Mistook His Wife for a Hat* if you want to learn about a man whose brain, for lack of a certain nutrient, can't form new memories. Then there's the woman whose brain doesn't quite assign the right "space" to her body parts.

Clearly the nuts-and-bolts machinery of the brain is behind
our perception of the outside world. But is it completely responsible for the perception of the *inside* world as well? Is there some
part of the brain that gives us the experience of consciousness
just as there is some part that gives us the experience of vision?
Is consciousness merely an evolutionarily advantageous way
of interacting with our environment so that we can help the
Universe be even lazier? And if so, does that mean we have no
more control over our "selves" than we do over how we process a
visual image?

One overarching problem in dealing with these questions
is that we're trying to use the machinery itself to understand
the machinery. In Kurt Gödel's world (you remember—the guy
who proved that time doesn't exist) that was a big no-no. Mathematically, if you want to prove a statement, you need a system
of mathematics that's bigger than the statement. For instance,
you can't show that 1 + 1 = 2 unless you have some general
rules about arithmetic in place already, rules that apply to more
than just that statement. It's an idea that is known as Gödel's
Incompleteness Theorem, and it's been applied (some would say
hijacked) by philosophers to claim that we are incapable of fully
understanding consciousness. To understand the conscious mind,
they argue, you need some kind of framework that is bigger than
the mind, beyond the conscious mind. That's a rather tall order
because, well, we're kind of stuck inside it.

It might well be that there is a "consciousness" center of
the brain, giving us the illusion that we have power over our
thoughts and deeds. Or perhaps it's just the *interactions* of the
100 billion neurons and the trillions of "support cells" that give
us that impression. If so, then Aristotle was qualitatively right
(again), even if he didn't have the right organ.

Or perhaps it's something more, something for which you'll
need to hunt down some books on philosophy or theology
because I can't possibly do the subject justice here.

However it came about, this consciousness has allowed us to be aware of our own interactions with the world. Consciousness drives us to be explorers, to interact with as much as we can, and to take in as many experiences as possible. We *want* to know what is under that rock. Maybe it's edible, after all. We *want* to see what happens when we push this button (especially if we are told not to). We are curious creatures, and we revel in our curiosity. We touch and feel and smell and hear and taste, and then our brains assimilate those experiences and invent ways to experience new sensations. We strive to know what other people are thinking, and then we communicate with other humans in ways that no other animals have mastered. We push limits of our experience in a way that *Prochlorococcus Marinus* just doesn't.

In short, we wonder.

7.4 It's a Wonder-Full Life

Whether it's an inevitable by-product of the development of a complex brain, a naturally selected, evolutionarily advantageous trait, or something that points to a different, unphysical realm, our ability to wonder is still wondrous. In a very real sense, we are what the Sumerians intuited: we are the Universe thinking about itself, wondering about our origins and our destiny. We are life, physical beings born from matter that was once light in the nascent Universe, and we feed on the light squeezed from our nearby star. Our minds can simultaneously grasp the vastness of space and the diminutive realm of subatomic particles. We really are amazing things living in a wondrous Universe. But perhaps, with bills to pay and deadlines to meet and laundry to do and a boss or teacher or spouse or child who's been a complete . . . um . . . yeah . . . you've become immune to all the wonders around you.

If that's true, then you need a jump-start, not just because this

Universe really is an incredible place, but also because wondering is actually good for you. It's fat-free, high in fiber, and, even more important, it has been shown that people who are inquisitive about the world around them tend to be more creative, more appreciative of life, and less fearful of new situations.

Wondering people are simply happier people.

Being new to this planet, children are natural wonderers. Just watch a baby for a while. You can almost see him thinking, 'I wonder how it would feel and taste if I chewed on those things?' Imagine how amazing it must be to realize for the first time that those fingers are actually attached to you, that "you" have extensions into the physical world. These discoveries are a delight for infants, and the laughter bubbles from them. Sometimes with real bubbles. Discovering the outside world is what drives children to do many of the things they do, and probably a good 95% of the things that drive parents crazy. Then comes The Question: "Why?" Try not to brush it off next time you hear it. It's the audible expression of the Universe's desire to understand itself. You wouldn't want to thwart the Universe's main desire, would you?

Unfortunately, various factors—not just distracted parents—can squish a child's curiosity like a prehistoric "slugster," slowly morphing them into curiosity-free adults who honestly believe that the best years of their lives were spent in high school. Fortunately, that can be at least partially remedied if only they can start wondering again.

"But what should I wonder about?" you might ask after a month of days that make the Sisyphean labor of endless rock-rolling look pretty inviting.

Oh, I don't know. Perhaps night, light, stuff, gravity, time, home, and wonder? Or perhaps why the sky is blue, or what some of that background static is on your TV? Perhaps . . .

"Those are too deep for me this evening. I'm in the market

for something more bite-sized. Some smaller wonders," you say. "After all, I've got to make dinner and then take out the trash."

Fine. Anticipating your request, I asked folks from all ages and all walks of life what they wonder about or even what simply amazes them about life. A brief sampling of their responses has been compiled below. Roll them around in your head, savoring their wonder. Pick one each day or each week and look at the Universe as a curious child does. And as you ponder them, you'll likely come to agree with the illustrator who said simply,

"Everywhere I look there are wonders. And so little time."

SMALL WONDER

Small Wonders

The following are real wonders provided by real people, only a few of whom were promised free copies of this book. If you have your own to add, feel free to submit your suggestion to www.sevenwondersoftheuniverse.com. Who knows? Maybe yours will appear in *Seven MORE Wonders of the Universe.*

❧ *I wonder about knowledge. All the stuff there is to learn, and all the stuff I have yet to learn.*

❧ *I wonder if I will ever get around to making all those recipes in my cookbooks.*

❧ *I wonder about dreams—are they random bits of scenery created by the sleeping brain firing synapses or something else?*

❧ *I wonder about our sense of humor.*

❧ *What makes a baby smile for the first time? What are they thinking or experiencing when they give that first, great big, gummy smile?*

❧ *I wonder about the perfect balance and reliability of nature. If the setting of the human play was one hundredth as unstable and unpredictable as the plot, we would go crazy.*

❧ *Why does light travel at the speed it does?*

❧ *Why are some people left-handed and some people right-handed?*

❧ *Could I have been anyone other than me?*

❧ *I think it's wondrous how the mind craves continuity and will create it, even when it's not there, like in optical illusions.*

❧ *I still, over twenty years after the events, hold great awe and wonder at the miracle of the births of my two daughters—the changes in my body, the joy and the pain, then a tiny person so much a part of me and yet so individual enters the world.*

❧ *I love how many different varieties of food we can prepare, but I wonder who first decided that food needed to be cooked?*

❧ *I'm curious about pets. I wonder how we developed such affection for beings of an entirely different species, and who share our homes and become such an important part of our lives.*

❧ *I wonder about friends and family and the unexpected kindness of a stranger.*

❧ *Why does the inside of an Oreo taste fine alone, but the outside only tastes good dunked in milk?*

❦ *I find music wondrous. It can convey every emotion there is and evoke so many different human responses without a single word ever being uttered.*

❦ *What makes a germinated seed of a vine search for a host tree and to climb up that tree toward the sunlight? Why does a chick peck at the shell until it is freed from its confinement? What makes newly hatched turtles instinctively dig their way out of the sand and head toward the hitherto unknown ocean at maximum speed?*

❦ *Why are some people ticklish and other people aren't?*

❦ *Walking on the beach is a wonder. All the little shells on the seashore that once held life. Even the broken ones are beautiful, remnants of lives that were so different from mine.*

❦ *I wonder what kind of bird is making that sound. No, not that sound. That sound.*

❦ *Butterflies and dragonflies and all the things that swim through the air are wondrous. I wonder what the world looks like through their eyes?*

❦ *Why do people always want to put things in order or classify things into categories?*

❦ *Where does all the lifetime's vast accumulated knowledge go to when you die? What about the emotional bank that stores the love you hold for someone loved? What happens to that when you have gone?*

❦ *I wonder what causes people to be musical or artistic or mathematical geniuses?*

❦ *I wonder how people go about inventing new things? How can they look at an object and think to take it in a whole new direction?*

❧ *Something that makes me wonder is the magnificence of nature. The Grand Canyon, the Great Barrier Reef, the Nordic Fjords, Niagara Falls, and the Antarctic, they all strike me with wonder.*

Afterword

The inquisitive child is sleeping now, dreaming of new questions to fling at you tomorrow. Your living room is adorned by the remnants of a dozen interpretive dances and demonstrations.

That can all be tidied another time. Right now the trash can beckons.

Tonight as you slog through this least wondrous of chores, pause to experience the wonder of it all. Breathe in the oxygen that was assembled in stars now billions of years dead. Step deliberately on the seemingly solid crust of your wondrous home as you drag the heavy load, weighed down by the mysterious gravity, to the curb. Revel in the darkness of night, punctuated by the light from the unimaginably distant bright dots that share in your origins.

Then go back inside and rest. There are more wonders to be discovered tomorrow.

Particularly if the raccoons manage to pry open your trash can.

For Further Reading

Books

Some of the following are sources for the material in this book. Some of these are related things that got me thinking but that I didn't directly draw from. Some of these are things that I *want* to read but haven't yet.

St. Augustine. *Confessions*. Translated by Henry Chadwick. Oxford: Oxford University Press, 1991. This is a great read if you want to brush up on your classic theological literature. As for this work, I was particularly interested in his thoughts on "time," which are found in Book XI.

Bennett, Jeffrey. *Beyond UFOs: The Search for Extraterrestrial Life and Its Astonishing Implications for Our Future*. Princeton, N.J.: Princeton University Press, 2008. If you liked *Seven Wonders of the Universe*, try *Beyond UFOs*, which goes into much more detail on many of the ideas touched on in chapters 6 and 7.

Bondi, Hermann. *Relativity and Common Sense: A New Approach to Einstein*. New York: Dover Publications, Inc., 1980. Just what it says it is. According to the history of science student who gave me this book, "It actually makes sense!" For relativity, that is.

Bryson, Bill. *A Short History of Nearly Everything*. New York: Broadway Books, 2004. Again, exactly what it says it is. This book is loads of fun but should be taken in small portions, like a rich cheesecake. Otherwise you might OD on scientific facts.

Danielson, Dennis R., ed. *The Book of the Cosmos: Imagining the Universe from Heraclitus to Hawking*. Cambridge, Mass.: Perseus Books Group, 2000. This is a really cool collection of 85 essays by the original authors (from Heraclitus to Hawking, in fact), with some helpful editorial/explanatory comments.

Harrison, Edward. *Darkness at Night: A Riddle of the Universe*. Cambridge, Mass.: Harvard University Press, 1987. This is *the* definitive work that compiles all the history and cosmology of the problem of the darkness of night.

Hawking, Stephen. *A Brief History of Time: From the Big Bang to Black Holes.* New York: Bantam Books, 1988. 'nuff said.

Hofstadter, Douglas. *Gödel, Escher and Bach: An Eternal Golden Braid.* New York: Basic Books, 1979. Mind-boggling. Complex and rich and beautiful—and I couldn't find a single thing in it that I could use in this book. But you should at least flip through it on your next trip to the bookstore.

Metzinger, Thomas. *Neural Correlates of Consciousness: Empirical and Conceptual Questions.* Cambridge, Mass.: MIT Press, 2000. For those of you who are deep into the biological and philosophical questions about consciousness (or want to be), this is your book.

Sacks, Oliver. *The Man Who Mistook His Wife for a Hat.* New York: Summit Books, 1985. If any book can make you extremely protective about your mental capacities, this is it. It's a collection of bizarre mental illnesses that you hope you will never suffer from. What's even more bizarre is that this has been adapted as an opera. Weird . . .

Schrödinger, Erwin. *What Is Life? with "Mind and Matter" and "Autobiographical Sketches."* Cambridge: Cambridge University Press, 1992. This is a compilation of the 1944 classic *What Is Life?* along with other works by the man who imagined placing a kitty with a radioactive substance.

Shubin, Neil. *Your Inner Fish: A Journey into the 3.5 Billion Year History of the Human Body.* New York: Pantheon Books, 2008. If you really want to know about DNA, evolution, the *Sonic Hedgehog* gene, and why you have hiccups and hernias, you should check out this tome. Very readable account by the fellow who found our fishy missing link.

Yourgrau, Palle. *A World without Time: The Forgotten Legacy of Einstein and Gödel.* Cambridge, Mass.: Basic Books, 2005. If you weren't completely convinced that Gödel was both a genius and a madman, you should read this.

Journal Articles

These are things you probably don't want to mess with unless you're pretty knowledgeable on the subject matter.

Arpino, M., and Scardigli, F. 2003. *European Journal of Physics.* 24: 39–45. "Inferences from the Dark Sky: Olbers' Paradox Revisited." A great read, assuming you speak math and are comfortable with phrases like "classical euclidean-newtonian manifold." But it does comment on conclusions presented in Harrison's book (see above).

Deines, S.D., and Williams, C.A. 2007. *The Astronomical Journal* 134: 64–70. "Time Dilation and the Length of the Second: Why Time-scales Diverge." See comments about the Arpino and Scardigli paper.

Gallagher, M.W., and Lopez, S.J. 2007. *The Journal of Positive Psychology* 2 4: 236–248. "Curiosity and Well-Being." A pretty straightforward article about how your curiosity about the world is related to your happiness. You're obviously happy, though, because you now appreciate the simplest wonders of the Universe and will no doubt go find out more about them.

Lazcano, A., and Miller, S.L. 1996. *Cell* 85: 793–798. "The Origin and Early Evolution Review of Life: Prebiotic Chemistry, the Pre-RNA World, and Time." Again, this is pitched at a scientific audience, but it distills decades of ideas on the origin of life into a single paper. And unlike *this* book, it does mention DNA.

Nagel, T. 1974. *Philosophical Review* LXXXIII 4: 435–450. "What Is It Like to Be a Bat?" If you're wanting a philosophical discussion of consciousness, this one is actually one of the most lucid. And who could resist reading something with that title?

Pereto, J. 2005. *International Microbiology* 8: 23–31. "Controversies on the Origin of Life." Another good, but scientific, review, and it has a whopping big reading list for those interested in getting to the original sources.

Rampino, M.R., and Caldeira, K. 1994. *Annual Review of Astronomy and Astrophysics.* 32: 83–114. "The Goldilocks Problem: Climatic Evolution and Long-Term Habitability of Terrestrial Planets." This is a great, relatively readable review article that summarizes what was known as of 1994 in the so-called Goldilocks problem, which is why Goldilocks features so prominently in this book.

Miscellaneous Virtual References and Other Neat Websites

Warning: Some of these websites might not exist at the time you read this. Check www.sevenwondersoftheuniverse.com for updates to this list.

The official site of the International Earth Rotation Service, for all your Earth rotation needs:

www.iers.org/

A pretty thorough site about stars and habitable planets. So thorough, in fact, that the site is called "Stars and Habitable Planets." It has the decency (unlike so many websites) to link to the original research papers on the topics discussed:

www.solstation.com/habitable.htm

Oodles of really fascinating information about life on Earth, the possibilities of life on other planets, important astronomical findings that help us understand life's beginnings, etc., can be found at:

www.astrobio.net

A relatively short and easy-to-read article about the origin of the Moon and the length of the day:

http://space.newscientist.com/article/mg19826525.500-moons-birth-changed-the-length-of-days-on-earth.html

What the night-sky view would be from a globular cluster:
   . . . as described in a science class at the University of Oslo:

www.uio.no/studier/emner/matnat/astro/AST1100/h04/undervisningsmateriale/lectures/lecture-17.pdf

   . . . and as described by several professional astronomers at the 1998 Canary Islands Winter School of Astrophysics:

www.iac.es/gabinete/iacnoticias/winter98/xplaneta.htm

   . . . and as visualized in a great music video that features Carl Sagan and Stephen Hawking:

www.symphonyofscience.com/ (scroll down to "A Glorious Dawn")

More about cosmic background radiation on television:

www.nasa.gov/vision/universe/starsgalaxies/cobe_background.html

A college classroom capsule summary of Big Bang chronology:

http://astro.ucla.edu/~wright/BBhistory.htm

A relatively down-to-earth article describing the B-sub-s quark experiment that might or might not explain why matter "won" in the Early Universe:

www.sciam.com/article.cfm?id=matter-antimatter-split-hi

Lord Kelvin tells us how it is in his classic paper about the Sun's energy source and expected lifetime:

http://zapatopi.net/kelvin/papers/on_the_age_of_the_suns_heat.html#fnsectii

Fun stuff about the interior of Earth:

> http://pubs.usgs.gov/gip/dynamic/understanding.html and http://earthquake.usgs.gov/research/structure/crust/index.php

What the Sun is doing at this very moment, and whether you'll be able to see any good aurora anytime soon:

> http://spaceweather.com

How we got our magnetic field. Maybe:

> http://es.ucsc.edu/~glatz/geodynamo.html

Lots of great information about the origins of life, sponsored in part by the National Science Foundation:

> http://exploringorigins.org/

Creepy things at the bottom of the ocean, and what *you* can do to help them stay creepy and at the bottom of the ocean:

> http://coml.org/

Another great music video by Symphony of Science with Carl Sagan, Richard Feynman, Neil DeGrasse Tyson, and Bill Nye about how we are all connected to each other and to the cosmos:

> www.symphonyofscience.com (scroll down to "We Are All Connected")

Why you should care about wondering:

> http://psychologytoday.com/articles/pto-20060831-000002.html

Index

The letter f following a page number denotes an illustration

aliens, 195–96

amino acids, 204–5

angular momentum. *See* conservation of angular momentum

antimatter, 97–101, 98f

ant on a balloon, 34, 36f, 66f, 69, 130

Aristotle: and consciousness, 215, 217–18, 221; and Greek elements, 107–8; and natural tendencies of objects, 107–8; and time, 147

arrow of time, 154–57. *See also* entropy

atom: constituent particles of, 77; number of, in human body, 4, 102; scale of, 77, 78f. *See also* elements

Augustine. *See* St. Augustine

aurora, 173

Babylonians, and timekeeping, 139, 141, 146

Big Bang, 69, 96, 199, 200f

black holes, 152f; falling into, 164–65; time inside, 153, 164

blueshift, 65

B_s meson, 100

c. *See* speed of light

caloric rays. *See* infrared light

Cavendish experiment, 111–12

cell structure, 211

Census of Marine Life, 195

Cesium-133, 136, 137, 151

CFL. *See* compact fluorescent light bulbs

charges: attraction and repulsion of, 80–84, 82f; bad joke about, 81; number of, acquired by balloon rubbed on hair, 109

Chronology Protection Conjecture, 161, 162f

closed time-like paths, 161

comets: Hyakutake, 183–84; and life on Earth, 205; naming of, 183–84; Shoemaker-Levy 9, 123; Wild 2, 205

compact fluorescent light bulbs, 45–46

Comte, Auguste, 67

consciousness, 215–22, 216f; and ancient Sumerians, 216f, 217; and Aristotle, 215, 217–18; and Plato, 217

conservation of angular momentum, 13, 14f, 15

continental drift. *See* plate tectonics

convection, 189–90, 192f

Cosmic Microwave Background Radiation, 37, 72, 96; detected on television, 75

crust: on Earth, 188–90, 192f; on Mars, 177–78

day: length of, in future, 19; length of, in past, 17, 19; measurement of, 137, 138; Spanish names of, 163

Daylight Saving Time, 161
Descartes, René, 159
disorder: and life, 202-4; tendency toward, 155-58
Doppler effect, 64-65, 66f

E = mc². *See* Einstein, Albert
Earth: crust of, 188, 192f; distance of, from Sun, 173; how far it will move if everyone in China jumps off a chair at the same time, 116; interior of, 190; magnetic field of, 190-91, 192f; mass of, 95, 107; radius of, 113; rotation rate of, 8-9, 19-21; your weight on, 112, 115, 133
Einstein, Albert: and E = mc², 88f, 94-96; and particle nature of light, 68; on time, 158. *See also* General Theory of Relativity
electrical charge. *See* charges
electricity, static. *See* static electricity
electrolysis, 92
electromagnetic spectrum, 61. *See also* infrared light; microwaves; ultraviolet light; x-rays
electron, 77, 80; number acquired by balloon rubbed on hair, 109; wave nature of, 84-85, 86f
electrostatic repulsion, 80-84. *See also* static electricity
elements: defining characteristic, 79; naturally occurring, 185; present in human body, 79-80, 101-2; present in the early Universe, 96, 185; spectra of, 49-50
energy-mass equivalence. *See* mass-energy equivalence
entropy, 155-58, 202-4; reversal of, 156f, 158. *See also* arrow of time
evil twins, 97-101, 98f
evolution: of atmospheres, 174f, 175, 177-78; of life, 209, 211-13; of ribozymes, 209

excuses for missing work, 61, 91
exoplanets, 22-23, 193-94
expansion of the Universe, 33-37, 69-72; and the arrow of time, 157-58; and the darkness of night, 35, 37; and its temperature, 92-95, 94f
Exploring Life's Origins (website), 209
extinctions, 213

fabric of spacetime. *See* spacetime
"facts," scientific, 80
Fermi National Accelerator Lab, 100
Feynman, Richard, 136
Flutie the fat cat, 132f
forces, fundamental, 109
Franklin, Benjamin, 80
free-fall, 116-19, 120. *See also* orbits
frictional braking, 19
fundamental forces, 109
fusion. *See* nuclear fusion

galaxies, 24. *See also* Milky Way
Galileo, 19, 30; and speed of light, 67
General Theory of Relativity, 33; and black holes, 153; and Global Positioning System satellites, 151; and Mercury's orbit, 127-29; as mimicked by gravity well, 131; and time, 148, 151, 159-60
George. *See* Uranus
Gliese 581: mass of, 179; planetary system around, 22-23, 179; temperature of, 179
Global Positioning System, 150f, 151, 153
globular cluster, 25; M13, 196. *See also* Palomar 4
glycine, 205. *See also* amino acids
Gödel, Kurt: and consciousness, 221; Incompleteness Theorem of, 221; and time, 159-60,
gossip: game of, 155; light as the cosmic, 68-69

GPS. *See* Global Positioning System
gravitational lensing, 129–30
greenhouse effect, 170–71; and plate
tectonics, 189

habitable zone, 194
hammer and feather experiment,
105f, 106
Hawking, Stephen, 161, 162f
Herschel, William, 52–53. *See also*
infrared light
Hilton, Paris, 190, 196
Hinode x-ray telescope, 59
hour(s): and the Moon, 141–45;
number of, in a day, 139
Hubble, Edwin, 71
human body: amino acids in, 205;
composition of, 79–80, 101–2, 202;
mass of, compared to Earth, 115;
number of atoms in, 102
hydrogen, 79, 96

ice-skaters, 13, 15, 19
IERS. See International Earth
Rotation and Reference Systems
Service
infrared light, 53–56, 54f; discovery
of, 53; and the greenhouse effect,
171, 172f; and temperature, 53,
55; wave nature of, 53. *See also*
William Herschel; Spitzer Space
Telescope
International Dark-Sky Association,
40–41
International Earth Rotation and
Reference Systems Service, 17,
18f, 146
International Occultation Timing
Association, 144
interpretive dance, 6, 15, 127, 142f,
143, 191, 203
interval. *See* time
IOTA. *See* International Occultation
Timing Association

Jupiter: mass of, 11; radius of, 11;
rotation rate of, 10–11; and Thursday, 163; your weight on, 133

Kelvin, Lord, 31; and lifetime of Sun
and stars, 124; and Sun's energy,
123–24; and temperature scale, 70
Kepler, Johannes, 30
Kepler Mission, 193–94

Large Hadron Collider, 131
lava lamp, 189, 192f
Law of Universal Gravitation.
See Newton's Law of Universal
Gravitation
leap seconds, 17
Le Verrier, Urbain, 125
lies your teacher told you, 80
life: emergence of, 211; and entropy,
202–5, 212, 213; evolution of,
211–13; and extinctions, 213;
ingredients for, 202; *What Is Life?*
(Schrödinger), 201
light: bent by gravity, 129–30;
blocked by atmosphere, 55, 57,
59; definition of, 63; scattering of,
72–74, 73f; speed of, 27, 67, 148;
stretching with Universe, 70; and
temperature, 45, 55; transmitted
by atmosphere, 59; travel time of,
31; types of, 52–63; wavelengths
of, 64–65; wave nature of, 53,
63–64. *See also* electromagnetic
spectrum; infrared light; microwaves; ultraviolet light; x-rays

magnetic field: on Earth, 191, 192f;
how planets create, 191, 192f, 193;
how to create using interpretive
dance, 191; on Mars 177–78, 193;
on Venus, 171, 173, 174f, 193
*The Man Who Mistook His Wife for a
Hat* (Sacks), 219
Mars: atmosphere of, 177–78; crust of,
177–78, 193; distance of, from Sun,

Mars (*cont.*)
175; gravity of, 178; magnetic field of, 177–78; and Martians, 176–77; mass of, 177; radius of, 177; rotation rate of, 9, 177; temperature of, 178; and Tuesday, 163; water on, 177, 178; your weight on, 133

mass-energy equivalence, 94, 95

Maxwell, James Clerk, 67

Maxwell's equations, 67

mean ol' acid. *See* amino acid

Mercury: day-night cycle of, 38–39; distance of, from Sun, 170; orbital peculiarities and General Theory of Relativity, 125, 129; rotation rate of, 9, 38–40; temperature of, 170; and tidal locking, 38; and Wednesday, 163

meson. See B_s meson

Michell, John, 111–12

Microwaves, 59, 61, 72, 167, 199

Microwave Background Radiation. *See* Cosmic Microwave Background Radiation

Milky Way: candy bar, 24; central black hole of, 27–28; location of Sun in, 24, 27; number of stars in, 24; position of, in Universe, 71–72; shape of, 25

Moon: distance of, from Earth, 121; hammer and feather experiment on, 106; and the hour, 143–45; mass of, 20, 112; and Monday, 163; and the month, 143; motion of, against background stars, 144; occultation by, 144–45; orbital period of, 20; radius of, 113; and tidal forces, 19; your weight on, 112, 133

monosodium glutamate, 205

muons, 148–49

Nagel, Thomas, 218

Neptune, 16; discovery of, 121–22; your weight on, 134

neutron, 77; and quarks, 94

Newton, Isaac, 30; and calculus, 113; and light, 63; and thought experiment about orbits, 116; and time, 146–47

Newton's Law of Universal Gravitation, 108–11, 113, 115, 124; failures of, 125; and tides, 122–23

Night (as seen from different locations), 23, 25, 27–30, 33–34

nuclear fusion, 82f, 83–84, 89–91, 99

Olbers' Paradox, 30

Oort Cloud, 183, 185

optical illusions, 219

orbits: and free-fall, 116–19, 118f; record-holder, 120; and spacetime, 131; speed required for, 119

Ørsted, Hans Christian, 191

ozone layer, 175

PAH. *See* polycyclic aromatic hydrocarbon

Pal 4. *See* Palomar 4

Palomar 4: distance to, 121; location of, 25, 26f, 28; view from, 25

paradox, 30

particle accelerators, 94, 100

penguins, 44f, 74f, 92, 93f, 200f, 210f

Perelandra (Lewis), 169

phase changes, 92

photosynthesis, 175, 205, 208

planetary formation, 13; and gravity, 123; timescales, 13, 123

planetary transits, 194

planets around other stars. *See* exoplanets

plate tectonics, 188–90; contribution of, to greenhouse effect, 189; and lava lamps, 189, 192f

Plato, 217

Pluto, 121

Poe, Edgar Allan, 31, 32f. *See also* Olbers' Paradox

Polyakov, Valerie, 120

polycyclic aromatic hydrocarbon, 204
Prochlorococcus Marinus, 207–8
proton, 77, 79; maximum temperature of, 94; and quarks, 94
Proxima Centauri, 27

quantum mechanics, 201
quark, 94, 100

Redi, Francesco, 207
redshift: from expansion of Universe, 71; from relative motion, 65. *See also* Doppler effect
relativity. *See* General Theory of Relativity
Rhyme, NoOne, 7f, 8, 117, 118f, 119
ribozyme, 208–9, 211
Ritter, Johann, 57
runaway greenhouse effect. *See* greenhouse effect
Ryan, Nolan, 5. *See also* Rhyme, NoOne

Sacks, Oliver, 219
Sagan, Carl, 76, 202
Saturn: mass of, 11; radius of, 11; rotation rate of, 11; and Saturday, 163; your weight on, 134
scale: of atoms, 77, 78f; of stars near the Sun, 24; of visible light within electromagnetic spectrum, 61
Schrödinger, Erwin, 201
Schrödinger's cat, 201
Scott, Commander David, 106
Search for Extraterrestrial Intelligence, 196–97
second, 136, 145
SETI. *See* Search for Extraterrestrial Intelligence
Slipher, Vesto, 71
SOHO. *See* Solar and Heliospheric Observatory
Solar and Heliospheric Observatory, 58

solar system formation: and gravity, 123; and static electricity, 13; and timescales, 13, 123, 131; and water, 179
spacetime: and gravitational lensing, 130; and the passage of time, 149, 151, 153; and redshifts, 70–71
spectroscope, 47–49, 50f
spectrum: and composition, 51; of incandescent bulb, 48; of "neon" signs, 49, 51; of stars and nebulae, 52; of Sun, 47; and temperature, 51
speed of light: 27, 95; Galileo's attempt to measure, 67; and passage of time, 148–49
speed of observers at planetary equators, 8–9
Spitzer Space Telescope, 56
spontaneous generation, 207
St. Augustine, 147–48, 154
stars: buying, 183–84; density of, 24, 26, 27; distribution of, in Milky Way, 25, 26f; formation of, 123; lifetime as computed by Lord Kelvin, 31; lifetime of a high-mass, 179–80; lifetime of low-mass, 180; number of, visible to naked eye, 24; source of energy of, 123–24; suitable as life-bearing planets, 179, 181–82, 187
static electricity: and solar system formation, 13, 81; versus gravity, 109
strong nuclear force, 82f, 83, 109, 110f
Sumerians: and consciousness, 216f, 217; and timekeeping, 139
Sun: as seen from Mercury, 38, 39; as seen in different wavelengths, 58–59; density of, 87, 89; distance of, from Earth, 173; formation of, 123; fusion within, 87, 88f, 89–90, 99; lifetime remaining of, 99; location of, within Milky Way, 26f; and

Sun (*cont.*)
 Lord Kelvin, 31, 99, 124; mass loss
  rate of, 99; orbit of, within Milky
  Way, 28, 182; rotation rate of, 16;
  size of, 8; as source of energy, 87,
  88f, 89–90, 123–24; space weather
  from, 173; and Sunday, 163; temperature structure of, 87, 89; tidal
  forces from, 21
sunsets, 72–75, 74f

television static, 75
things that will kill you: achiev-
  ing minimum thermodynamic
  energy, 205; asteroids and comets,
  181, 184; being in an infinitely
  big, old Universe, 30; being too
  close to the Sun, 21; black holes,
  164–65; extreme tidal forces,
  164; fast-moving electrons, 173;
  fast-moving protons, 173; gamma
  rays, 165; living near the Galactic
  center, 27–28; massive glaciation,
  213; Venus, 9, 169; violent solar
  flares, 179; wormholes, 164–65;
  x-rays, 59
Thomson, William. *See* Kelvin, Lord
tidal forces, 19–21, 122–23; and
  black holes, 153, 164
tidal locking, 20–21; and the Gliese
  581 system, 22; and Mercury, 38
tides. *See* tidal forces
time, 3, 123, 135–64, 138f, 162f; and
  Isaac Newton, 146–47
time dilation, 148–49; and GPS, 151

ultraviolet light: discovery of, 57;
  and ozone, 175; and temperature,
  58
units of time: and ancient Babylonians, 139; and ancient Sumer-

ians, 139–41; day, 137; hour, 141;
  second, 136; week, 161, 163
Universe: age of, 35–37; composition of, 96; energy in early, 95, 96;
  expansion of, 33–35, 36f, 69–72;
  and laziness, 202–4; mass of, 95;
  origin of, 35, 37; size of, 70; temperature of, 37, 70; what God was
  doing before He created, 147
Uranus, 16, 52, 121

Venus: atmosphere of, 169, 174f,
  175; cloud cover of, 167, 174f;
  compared to Earth, 174f; distance
  of, to Sun, 170; and Friday, 163;
  greenhouse effect of, 171, 174f;
  historical views about, 167, 169;
  magnetic field of, 171, 173, 174f;
  nicknames for, 167; rotation rate
  of, 9, 167, 169; spacecraft on, 170;
  temperature of, 169; water loss of,
  170, 174f; your weight on, 133

waves, 63–65, 68; and electrons,
  84–85, 86f
weak nuclear force, 109
week, length of, 161, 163
weight: on Ceres, 133; on Earth, 112,
  115, 133; and General Theory of
  Relativity, 128; on Jupiter, 133; on
  Mars, 133; on the Moon, 112, 133;
  on Neptune, 134; on Saturn, 134;
  on Venus, 133
"What is it like to be a bat?" 218
Wheeler, John Archibald, 136
Wollaston, Francis, 111–12
wormhole, 164–65

x-rays, 59

zero g, 128. *See also* free-fall; orbits